D0428914

WIND WIZARD

WIND WIZARD

Alan G. Davenport
and the Art of Wind Engineering

SIOBHAN ROBERTS

PRINCETON UNIVERSITY PRESS

PRINCETON AND OXFORD

Copyright © 2013 by Siobhan Roberts
Requests for permission to reproduce material from this work should be
sent to Permissions, Princeton University Press
Published by Princeton University Press, 41 William Street, Princeton,
New Jersey 08540
In the United Kingdom: Princeton University Press, 6 Oxford Street,
Woodstock, Oxfordshire OX20 1TW

press.princeton.edu

Jacket photograph by Ron Nelson. Courtesy of the estate of Ron Nelson.

Library of Congress Cataloging-in-Publication Data
Roberts, Siobhan.
Wind wizard : Alan G. Davenport and the art of wind engineering /
Siobhan Roberts.
p. cm.
Includes bibliographical references and index.
ISBN 978-0-691-15153-3 (hardcover : alk. paper)   1. Wind-pressure.
2. Davenport, Alan G.   3. Buildings—Aerodynamics.   4. Bridges—
Aerodynamics.   I. Title.
TA654.5.R636 2012
624.1′75—dc23   2012015170
British Library Cataloging-in-Publication Data is available
This book has been composed in Minion Pro
Printed on acid-free paper. ∞
Printed in the United States of America
1 3 5 7 9 10 8 6 4 2

In memoriam

Alan Davenport

1932–2009

CONTENTS

WIND WIZARD

I

Sowing Wind Science

No sooner did the Tacoma Narrows Bridge—the world's third longest suspension bridge, and the pride of Washington State— open in July 1940 than it earned its epitaphic nickname, "Galloping Gertie." The 4,000-foot structure, its main span reaching 2,800 feet, twisted and bucked in the wind. The pronounced heave, or more technically speaking the longitudinal undulation, caused some automobile passengers to complain of seasickness during crossings. Others observed oncoming cars disappearing from sight as if traveling a hilly country road. By November 7, amid 39-mile-an-hour winds, the $6,400,000 bridge wobbled and flailed, then rippled and rolled, then twisted like a roller coaster, until in its final throes it plunged, with a beastly roar, 190 feet into the waters of Puget Sound.

See Glossary for definitions

Speaking to a *New York Times* reporter the day after the collapse, Leon S. Moisseff, the bridge's designer and engineer, was at a loss to explain the cause, placing blame on "a peculiar wind condition."

Wind engineer Alan G. Davenport, founder of the Boundary Layer Wind Tunnel Laboratory at the University of Western Ontario, often summoned the memory of the Tacoma Narrows Bridge disaster as a cautionary tale. "Most features of this disaster are too familiar to bear repeating," he told his audiences, whether assembled at technical lectures or at popular talks. Both occasions always included screenings of grainy film footage capturing the bridge misbehaving as though fashioned from rubber—footage now preserved, owing to its cultural, historical, and aesthetic import, in the United States National Film Registry, as well as on YouTube, with numerous clips garnering more than six million cumulative views. Nonetheless, Davenport noted, as familiar as this disaster may be to the collective consciousness, the

Figure 1A. The Tacoma Narrows Bridge displayed torsional oscillation and longitudinal undulation even before it opened on July 1, 1940. University of Washington Libraries, Special Collections, UW21413.

Figure 1B. Photographer Howard Clifford of the *Tacoma News Tribune* snapped a few shots and ran. University of Washington Libraries, Special Collections, UW20731.

Figure 1C. The bridge's main span collapsed into the waters of the Tacoma Narrows on November 7, 1940. University of Washington Libraries, Special Collections, UW21422.

Figure 1D. The new and improved Tacoma Narrows Bridge, 1950. Courtesy of R. A. Dorton.

consequences bear continued consideration. "What's past is pro-
logue," Shakespeare observes in *The Tempest* (and said the jester Trin-
culo, "Here's neither bush nor shrub to bear off any weather at all, and
another storm brewing; I hear it sing i' th' wind"). In broadening the
moral of the Tacoma disaster and applying it to the behavior of all
structures—bridges, buildings, and beyond—Davenport made of this
cautionary tale a professional credo that governed his lifelong fasci-
nation with the wind, and with balancing the wind's fickle forces.

It was in 1965 that Davenport established the world's first dedicated boundary layer wind tunnel designed to test civil engineering
structures. The planetary boundary layer is the region of the atmosphere extending from the earth's surface upward about 3,000 feet,
the wind churning the air into turbulent eddies, average velocity increasing with height. A boundary layer wind tunnel mimics these
marbled striations of air—it mimics wind energy—in order to test
designs for buildings and bridges that will face the wind when built.
Making its debut in the 1960s, Davenport's wind tunnel arrived the
same decade as the laser, the computer mouse and the Internet, hand-
held calculators and the ATM, Apollo 8, string theory, and Rachel
Carson's *Silent Spring*. In the years to come, Davenport's revolutionary lab would investigate the windworthiness of some of the world's
most innovative structures: many of the tallest buildings, including
New York City's World Trade Center and Citicorp Tower, Chicago's
Sears Tower, Boston's John Hancock Tower, Shanghai's World Financial Center, and Toronto's CN Tower (which, strictly speaking, is not
a building but rather a freestanding structure), and many of the longest bridges, among them Florida's Sunshine Skyway, the proposed
Straits of Messina span in Italy, France's Millau Viaduct and the Pont
de Normandie, as well as the iconic Golden Gate Bridge and New
York's Bronx-Whitestone Bridge. The Bronx-Whitestone Bridge,
being very similar in design to the Tacoma Narrows crossing, has
over the years required an extensive rehabilitation regime that continues to this day.

In addition to the buildings and bridges that came through the lab,
there were also a few exceptions and eccentricities. Legend has it that
in the early days the lab conducted tests on portable toilets, and later
on Arctic tents to be deployed by the Canadian military. NASA commissioned a study on the ground wind loads for the *Jupiter* launch

vehicle (occasionally Davenport said he wished he'd been an astronaut). The 2,421-foot illuminated Glorious Cross of Dozulé had its day in the tunnel, though it has yet to grace the countryside of Normandy. *Sports Illustrated* splurged on an investigation of Augusta National's twelfth hole, the lynchpin of the Masters' famed Amen Corner, said to be among the toughest holes in golf, in part because of the seemingly indecipherable winds (see sidebar, "Driving into the Wind," below).

Driving into the Wind

Sports Illustrated turned to the lab to decipher the maddening winds at what's been called the "meanest hole" in golf, the 12th hole at Augusta National. "Augusta National is a one-of-a-kind golf course," the article reported, "but all it takes to reproduce it (albeit at a scale of 1 to 200) is high-density foam sculpted with drywall compound, more than 600 trees made of sponge and wire, an acrylic Rae's Creek (complete with tiny silicone waves) and, for good measure, foam golfers that are nearly as stiff as the real thing. . . . The shot's path was represented on the model by a fixed piece of copper tubing 5/16th of an inch in diameter. Meteorological data from 1949 through '99 (collected at Augusta Regional Airport, about 10 miles south of Augusta National) was then analyzed by computer to create a simulation of the typical April winds that blow through Amen Corner. Smoke was used to give these breezes visual paths. To illustrate the turbulence at higher elevations, a wire coated with oil was fixed upwind from the model. An electrical current was sent through the wire until the oil burned, producing yellowish smoke. To depict the wind's effects along the trajectory of the shot, 13 evenly spaced holes were drilled along the copper tubing. Inside, titanium tetrachloride was introduced, producing bright white smoke." Results showed that one wind took two directions: "On the tee the wind is in the golfer's face, quartering slightly to the left (east), in the direction of the 11th fairway. About 25 yards into its flight the ball encounters a crosswind blowing to the east. Another 40 yards toward the green, as the shot is approaching its apex, the ball is slammed by a wind shear, with gusts blowing to the west toward the 13th fairway. This wind dissolves into low-speed swirling 20 yards from the green, as the ball is passing over Rae's Creek." The lab's project leader, Greg Kopp, concluded, "The challenge used to be trying to figure out the wind. Now the players have all the information, but they may wish they didn't. It's still a frightening shot into a very difficult wind."

Figure 2. Wind tunnel tests on Amen Corner, Augusta National Golf Course. Courtesy of Robert Walker for *Sports Illustrated*.

Figure 3. In the early days, the lab conducted tests on a design for portable toilets, and later on a ten-man Arctic tent to be deployed by the Canadian military. Courtesy of the Boundary Layer Wind Tunnel Laboratory.

Figure 4. Stabilizing the Glorious Cross of Dozulé against the wind in Normandy, France, proved difficult to finance (the clients attempted to get papal recognition of the location as the site of a miracle, which would have made raising the necessary funds easier). Courtesy of the Boundary Layer Wind Tunnel Laboratory.

Figure 5. Davenport with a model of the *Jupiter* launch vehicle. In 1966, he presented two papers at NASA's Langley Research Center for the "Meeting on Ground Wind Load Problems in Relation to Launch Vehicles." Courtesy of Alan Noon.

Himself a great sailor, Davenport reveled in the testing of sails for an America's Cup vessel. There were also investigations into how better to spray fruit trees so that the mist would not be blown off course by the wind. The solution proposed by Davenport's colleague, the electrical engineer Ion Inculet, was to make the fluid electrostatically charged, so that it would be attracted to trees and repelled from the ground. There were inquiries into how to ensure a clean airflow over surgical patients during hip replacement procedures, which are particularly susceptible to infection. And after the construction in 1984 of a second-generation wind tunnel that doubled as a wave tank, Davenport tracked the wind-induced drift of icebergs and observed the battering of BP and Exxon oil rigs in the open seas.

With such a multidisciplinary portfolio, Davenport quickly accumulated unparalleled expertise in the nascent field of wind engineering—indeed, the field emerged and evolved largely because of his work. With one pioneering example after another, he set the agenda for investigating the effects of wind on the natural and built environments.

Figure 6. In 1984, with the construction of a second-generation wind tunnel facility that doubled as a wave tank, the lab began studying wind and waves—investigating the effects of wind on oil rigs and icebergs. Courtesy of the Boundary Layer Wind Tunnel Laboratory.

Figure 7. Davenport and his daughter Clare sailed windsurfers in the wave tank at the opening of the new wind tunnel facility. Courtesy of the Boundary Layer Wind Tunnel Laboratory.

When the interaction between wind and our environs is not properly factored into structural design, the consequences can be catastrophic. A powerful lashing of wind assaulted a family viewing Christmas lights during a walk around Toronto's City Hall in 1982. The wind lifted the plywood from the promenade, threw the family into the air, over a protective parapet, and then dropped them 20 feet onto Nathan Phillips Square below, with serious injuries resulting. Climate change is arguably exacerbating extreme weather events, such as the deadly tornado outbreak in the southern United States in 2011, Hurricane Katrina in 2005, and the North American ice storm in 1998. And our exposure and risk are only heightened in the more fringe and fragile edges of nature where people are living, in both developed and developing countries—whether those fringes are the Florida Gulf resort communities or the rural coastal villages of Sri Lanka, where Davenport attended a disaster relief conference after the Boxing Day tsunami of 2004. Wind engineering is vitally concerned with how to prevent wind-induced disasters, the most costly disasters in terms of property damage and casualties. And although most knew him as a wind expert, Davenport in his versatility was among the first to advocate that the same mindset of preparedness be applied to all shapes and forms of natural disaster, not just those powered by wind.

Figure 8A. The lab put to the test several structures by Spanish architect Santiago Calatrava, including the Valencia Opera House. Courtesy of the Boundary Layer Wind Tunnel Laboratory.

Figure 8B. Calatrava's World Trade Center PATH Terminal at Ground Zero. Courtesy of Santiago Calatrava.

Upstream, Downstream

Pollution control efforts to cut sulfur emissions from chimneys at old coal-burning power stations resulted in a busy time for the lab's Barry Vickery, an expert on towers, chimneys, and stacks. (Vickery is now retired and serving as a consulting director of the lab.) In 1979, the Canada-USA Sulphur Emissions Reduction Protocol mandated a 30 percent reduction in emissions by 1993. To meet this mandate, companies such as American Electric Power, in the Ohio Valley, faced constructing new chimneys at preexisting plants with no choice but to put them in proximity to both the generating unit and the old chimneys. "Companies often want to avoid the considerable cost of dismantling the old chimneys," explains Vickery. "But having a number of stacks clustered together results in nasty aerodynamic problems—the upstream and downstream structures strengthen the vortex shedding produced by each, creating interference effects and loads powerful enough to damage these tall stacks, sending their exterior brick walls flying, or knocking the interior linings and filters." Vickery developed the best method for predicting the response of chimneys in such scenarios, a method he subsequently applied to well over 100 chimneys. Problems resulting from interference effects can be solved in a variety of ways, but simply removing the top third of an old stack due for decommissioning usually solves the problem.

Figure 9. The close proximity of cooling towers and old and new chimneys precipitates aerodynamic problems. Mark William Richardson/Shutterstock .com

In a career spanning half a century, Alan Davenport published hundreds of technical papers and scientific reports on the wind. He also seized every opportunity to impart a more popular, cultural, even philosophical and romantic perspective. One staple talk in his repertoire, "Sowing the Wind," he delivered at the 1979 commencement ceremony at Belgium's University of Louvain upon receiving one of his numerous honorary degrees. He explained that the title was taken from the words of the Old Testament prophet Hosea, "For they have sown the wind, and they shall reap the whirlwind," suggesting that one must heed the action and power of the wind or deal with the consequences. "Our ancestors, and civilizations before us in the ancient world, respected the wind for both practical and spiritual reasons," he observed. The wind has the power to drive turbines and evoke emotions. It pushes ships, and sinks them. It winnows grain and turns windmills grinding grain to flour, but it also flattens crops and blows down barns. Spiritually, Davenport noted, "Primitive people were no doubt awed by wind" . . .

> What exactly they thought about this mysterious invisible force we cannot be sure. Did they believe, like the popular cartoon philosopher Charlie Brown, that the clouds pushed the wind along? Or, like Ogden Nash, that the wind is caused by the trees shaking their branches? We can only guess that the unseen hand causing the rippling of the water and the rustling of the leaves was a sign that the natural world was animated by life-giving forces; that when these ripples turned to storm tossed waves, those forces were angry.

The principal god of the Aztecs, Davenport noted, was Quetzalcoatl, the plumed serpent and the god of wind. Quetzalcoatl's equivalent in ancient Egypt was Ammon, who shared the same godly status as Ra, the sun god. And the wind, in fact, derives its power not from the clouds or trees but from the sun—from the thermal shifts of the day-night cycle, as well as from the cooling effects of bodies of water and higher altitudes. Wind is defined by Environment Canada as the horizontal movement of air relative to the earth's surface caused by geographic variations in temperature and pressure. Solar radiation, beating down strongest at the equator and weakest at the poles, produces temperature and pressure differentials in the air. These pressure pockets are never stationary, always changing in their patterns, and altered by the earth's rotation. Air follows an eddying motion as it moves from areas of high pressure to low, rising as it warms, bringing cold air rushing in below to take its place. This mixing creates wind—the easterly trade winds of the Northern Hemisphere, which steer tropical storms across oceans and onto continents, and the violent prevailing westerlies of the Southern Hemisphere, known as the Roaring Forties, Furious Fifties, and Shrieking Sixties.

Figure 10. A few specimens from Davenport's collection of wind masks. Courtesy of the Boundary Layer Wind Tunnel Laboratory.

The wind was first treated as a subject of scientific inquiry in ancient Greece, and Aristotle's treatise on the elements, *Meteorologica*, written in 350 BC, endured as the standard reference in Europe for nearly two thousand years. Aristotle's pupil Theophrastus wrote two treatises on winds and weather, proposing that winds could be predicted according to the behavior of animals and the human body: "A dog rolling on the ground is a sign of a violent wind. . . . If the feet swell there will be a change to a south wind. This also sometimes indicates a hurricane." From Sir Isaac Newton, in his *Principia*, published in 1687, came the discovery that the wind force on a given shape is directly proportional to the shape's area, the air density, and the square of the wind's velocity. In 1759, another Englishman, John Smeaton, often called "the father of civil engineering," proposed a colloquial classification of the wind's force in a paper presented to the Royal Society. He listed eleven "common appellations," corresponding to the velocity of the wind by miles covered in one hour:

1	Hardly perceptible
2–3	Just perceptible
4–5	Gentle pleasant wind
10–15	Pleasant brisk gale
20–25	Very brisk
30–35	High winds
40–45	Very high
50	A storm or tempest
60	A great storm
80	A hurricane
100	A hurricane that tears up trees, carries buildings before it, etc.

In 1805, the Irish admiral Sir Francis Beaufort devised his eponymous Beaufort Wind Force Scale. It provided a standardized measure of wind speed based on sea conditions and is still used today (though now calculated with an empirical formula factoring in land conditions), ranging from zero, a calm and flat sea, to twelve, hurricane force with huge waves, foam, and spray.

By the end of the nineteenth century, fluid dynamics—the scientific domain with air and wind flow in its purview, as well as water flow—had advanced considerably. Equations developed in classical hydrodynamics produced beautiful solutions. The flow of air around a cylinder, for instance, produced a symmetrical pattern framed neatly by two stagnation points, the place where the air would come to a stop at the front and again at the back of the structure. But since classical hydrodynamics assumed ideal fluids—fluids with no viscosity, no resistance or friction—these beautiful solutions seemingly did not pertain to any real-life problems. As the British chemist and Nobel laureate Sir Cyril Hinshelwood reportedly lamented, "In the 19th century fluid dynamicists were divided into hydraulic engineers who observed things that could not be explained, and mathematicians who explained things that could not be observed."

In the early twentieth century, however, Ludwig Prandtl, a German scientist known for his seminal work applying mathematics to aeronautics, wedded these theoretical and practical solitudes. He proposed the concept of a boundary layer, and the notion that a viscous fluid actually possesses no velocity at the surface when it passes over an object. In dealing with ideal fluids mathematically and theoretically, as classical aerodynamicists did, fluids had been assumed to have no viscosity simply for the sake of simplicity, to make complex calculations easier. Now what Prandtl proposed was that this idealized theoretical assumption could be extended to practical applications as well. That the wind blowing over any surface doesn't have any velocity at the very interface with the surface is counterintuitive—a property of wind that even concrete-minded engineers describe as "magical," though it can be explained. If a viscous fluid such as air flows over a surface—say, the surface of a building, or the surface of the Arctic tundra—it is a fundamental property of the air that its particles have no motion when they meet the surface. Instead, on first contact a layer of air particles sticks to the surface. This is because at the microscopic level any surface is a craggy affair, and individual air particles can't help being trapped in the crags. Following this initial layer of air particles there is a second layer that viscously slides over the first. As a result, the second layer does have some velocity when it meets the surface; it shears past at a very slow velocity, dragged down by the particles that are locked into the crags. And so it goes, up and up and up. As the air moves away from the surface, its speed rapidly increases, and eventually the fluid reaches a speed more or less independent of viscosity, driven by a pressure field somewhere else, above or below. This stratification of the air and its variegated velocity make up the boundary layer.

While Prandtl applied his insights about the boundary layer mostly in the field of aeronautics, others, such as the French engineer Gustav Eiffel, played with structures of various shapes in wind tunnels. Eiffel

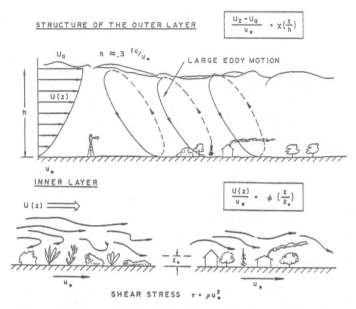

STRUCTURE OF THE OUTER LAYER $\qquad \boxed{\dfrac{U_z - U_G}{u_*} = \chi\left(\dfrac{z}{h}\right)}$

$U_G \qquad h \approx .3 \ ^{fc}/u_*$

LARGE EDDY MOTION

$U(z)$

h

u_*

INNER LAYER

$U(z) \Longrightarrow$

$\boxed{\dfrac{U(z)}{u_*} = \phi\left(\dfrac{z}{z_*}\right)}$

$u_* \qquad\qquad\qquad\qquad\qquad u_*$

SHEAR STRESS $\quad \tau = \rho u_*^2$

Figure 11. Davenport's illustration showing how large and small eddies make up the planetary boundary layer. Courtesy of the Boundary Layer Wind Tunnel Laboratory.

was one of the earliest engineers to factor in the important effects of wind on tall structures. When he designed his famous tower in 1887, he told the French newspaper *Le Temps* (defending himself against an artist's protest), "Now to what phenomenon did I give primary concern in designing the Tower? It was wind resistance. Well then! I hold that the curvature of the monument's four outer edges, which is as mathematical calculation dictated it should be ... will give a great impression of strength and beauty, for it will reveal to the eyes of the observer the boldness of the design as a whole." By Eiffel's account to the Société des Ingenieurs in 1885, however, his wind calculations had been cautiously conservative: "With regard to the exposed surfaces, we have not hesitated in assuming, in spite of the apparent severity of the assumption, that on the upper half of the tower all the lattice work is replaced by solid surfaces; that in the intermediate section, where the openings become more important, the frontal area is taken as four times the actual area of iron; below this (the first stage gallery and the upper part of the arcs of the legs) we assume the frontal area is solid; finally at the base of the tower we count the legs as solid and struck twice by the wind (i.e. each leg separately exposed to the full force of the wind)." Completed to mark the occasion of the Paris Exhibition in 1889, the Eiffel Tower became the world's tallest structure. At 986 feet, it almost doubled the height of the Washington Monument, its predecessor as the world's tallest (and the Eiffel Tower retained the title until surpassed in 1930 by New York's 1,050-foot Chrysler Building).

Soon after the tower's completion, Eiffel installed himself in a laboratory at the top level and began experiments on subjects of meteorological and engineering interest. He hung a vertical cable from the second level of his tower, along which he dropped various objects to test their drag. He later built a wind tunnel facility at the tower's base. And he measured the effect of the wind on the Eiffel Tower itself, using a vertical telescope positioned at the base, aimed at a target on his laboratory's underside, and correlating the deflections with measurements of wind speed at the peak. Exposed to high winds approaching diagonally, the tower was found to sway in an elliptical path with a transverse diameter of three to four inches.

Experiment and experience are the best teachers. Too often, though, modern success is blinding, and history's lessons are forgotten. This is the mantra of Henry Petroski, a professor of civil engineering at Duke University with a special interest in bridges and the author of (among a long list of books on the subject) *To Engineer Is Human: The Role of Failure in Successful Design*. With a photograph of the collapsed Tacoma Narrows Bridge on the book's cover, Petroski therein points to this structural failure as an unfortunate example of engineering hubris and oversight. The Tacoma Narrows Bridge's collapse, in 1940, followed half a century of relative calm after a string of bridge failures—the Menai Suspension Bridge, over the Menai Strait, North Wales, completed in 1826, was severely damaged by wind in 1839; the deck of Ohio's Wheeling Suspension Bridge collapsed in a windstorm in 1854, five years after being built; the Tay Rail Bridge crossing the Firth of Tay in Scotland collapsed in 1879, after less than two years in service; and the Niagara Falls Suspension Bridge, built by Samuel

Keefer, stood from 1869 to 1889, when it was felled in the night by a violent storm. These failures should have advanced the understanding of wind loading considerably and taught engineers to take heed. But as Petroski observed, the nineteenth-century failures were forgotten with the great successes of the Brooklyn Bridge, in 1883, and its structural descendants. Then, as engineers strove for ever-longer and lighter bridges, wooed by new theories, technologies, and innovative materials, their confidence was again upended by the extraordinary failure at Tacoma Narrows. "Certainly no designer who remembers the ill-fated Tacoma Narrows Bridge will design another bridge like it," Petroski says.

Hence Davenport, in his lectures on the past, present, and future of wind engineering, repeated the familiar features of the Tacoma Narrows Bridge disaster that should hardly have borne repeating—and, known for speaking at a calm, contemplative tempo, Davenport was never one to waste words. He also had the habit of sharing his inclination for learning lessons about the wind in unexpected and unusual places. "It is interesting to consider the effects of wind in nature," he once noted of an unlikely role model: "Palm trees, for example, exhibit a remarkable adaptation to resisting strong winds such as hurricanes. Over the millennia they have developed a remarkably tough, fibrous trunk with excellent fatigue resistance. By furling their branches, they reduce wind resistance so that their drag coefficient reduces dramatically with an increase in wind speed. This is a strategy not yet perfected by engineers," he said ruefully, adding that, "In contrast, humans confront the wind by leaning forward and relying on gravity!" Clearly a more comprehensive scientific approach was necessary. On that front Davenport forged forward with characteristic persistence and focus.

ॐ ॐ ॐ

The citation for his Canada Gold Medal in Science and Engineering noted, "It would be absurd to claim that Davenport's interest in airflow and wind behavior was innate. Yet it almost seems so." Herewith a brief recap of a wind wizard's formative years.

Alan Garnett Davenport was born in India's bustling city of Madras on September 19, 1932. His father, Thomas Davenport, of

Figure 12. Sandbags on rooftops warded off the wind at Davenport's childhood family estate in the Anamalai Hills of southwestern India. Courtesy of Sheila Davenport.

Cheshire, England, was a tea estate manager for the Scottish-owned London firm the Bombay Burmah Trading Company, situated in the Anamalai Hills of southwestern India. Traveling to the tea plantation involved a steep climb along a ghat, or pass, with forty-two hairpin bends. At the top, a 4,000-foot Shangri-La of bright green tea bushes emerged into view. Local denizens included elephants, tigers, black panthers, scorpions, cobras, porcupines, and monkeys. Ants were uninvited guests at the family home, a sprawling teak-paneled bungalow tended by his mother, Clara May Davenport (née Hope), along with her Indian staff.

The Davenport estate was called "Thoni Mudi," meaning boat-shaped hill. Perched at the top of the range, it caught exotic winds, winds from the cyclones and monsoons that swirled through the region seasonally. In 1935, a tropical cyclone hit India, killing sixty thousand people—one of the worst natural disasters in recorded history. To keep houses secured against the gusty winds, the general practice was to place large rocks and sandbags on the corrugated iron rooftops. Visiting Thoni Mudi with his brother Rodney and their wives fifty years later (after delivering the keynote address at the Indian National Structural Conference at Madras in 1993), Davenport was amazed to see how little had changed. Even the crude engineering precaution of rocks on roofs had endured.

Alan had two older brothers, Rodney, born in 1926, a historian living in Cape Town, and Tony, who was born in 1927 and knocked over and mortally injured by a truck in 1935. In October 1940, after the outbreak of World War II and their return with their parents to India from leave in England, Alan and Rodney were sent for safety's sake to live with relatives in South Africa. They traveled by ship, the fear of U-boats warranting a destroyer escort and a zigzag course to deceive the enemy. Alan found himself entertained in the evenings by the other children mounting dramatizations of dogfights using model airplanes and flashlights.

For the duration of the war and its aftermath, Alan was buffeted from school to school, guardian to guardian. He and Rodney lived with his mother's older brother, Arthur Hope, and his family for a time in Johannesburg. When Alan contracted tuberculosis he was sent to convalesce with his mother's younger brother, Ronald, who farmed in the Transvaal. The one constant he carried with him from place to place was his growing passion for aircraft. He not only made model airplanes, he designed them, carving the smaller craft from solid wood. He equipped the larger ones, as large as several feet in wingspan, with miniature gasoline engines not much bigger than spools of thread. Some put-putted around within a modest radius until they crashed, while others flew too far to recover (over half a mile). He acquired a love of aerodynamics and aeronautics at the tender age of fourteen, which comments on his report cards confirmed, noting he was, "A keen photographer and a brilliant aero-modeller." Comments also remarked on his capacity to lead. "He made a good prefect, reliable, conscientious and not afraid of responsibility. His handling of boys under his control was characterized by a quiet firmness which won him the respect of all. He is a thoughtful as well as an intelligent boy with an unusually developed sense of proportion and values . . . A very able and promising boy in every way."

Alan returned to England as a teenager, and since he was yet too young to attend university he spent a further year at the Repton School after having matriculated in South Africa. While studying for

Figure 13 (left). Alan Davenport as a youngster. Courtesy of Sheila Davenport. **Figure 14 (right).** Alan Davenport with his elder brother, Rodney. Courtesy of Sheila Davenport.

his entrance exams to Cambridge University he worked briefly for an actuarial firm, taking a cue from an aptitude test suggesting that statistics and numbers matched his inclinations. But then, inspired after hearing astronomer Fred Hoyle speak on the subject of the expanding universe, he took up mathematics in his first year at Cambridge, though he devoted more time to bridge, golf, and a cricket club that aimed to tie rather than win matches. In his second year he switched to the more practical mechanical sciences. Still, his extracurricular activities predominated, launching him onto his lifelong trajectory. Recruited as a cadet in the Royal Air Force, he flew between morning and afternoon lectures and made weekend jaunts in propeller planes over the English countryside, taxiing grassy airfields from Cornwall to Yorkshire. He also devoted considerable amounts of time as a writer and editor for *Light Blue*, the Cambridge sports magazine. During a reportorial visit to Oxford in 1954, he witnessed the first sub-four-minute mile, run by Roger Bannister. High winds on race

Figure 15. As a student at Cambridge's Emmanuel College, Davenport (in the plaid waistcoat) joined the Pagans' Cricket Club, whose mission was to play every match to a draw. Courtesy of Sheila Davenport.

day, at times reaching 25 miles per hour, had nearly convinced Bannister not to run. But the blustery conditions subsided as race time approached, and he ran the mile in 3:59.4 before a roaring crowd (with thousands more cheering in front of their television sets).

England's World War II rationing had only just ended when Davenport graduated with a bachelor's degree in engineering in 1954, so he set out for Canada to seek his fortune in a burgeoning economy. After botching an interview for a journalism job (confusing St. John's, Newfoundland, with Saint John, New Brunswick), he took a position as a lecturer teaching surveying at the University of Toronto. He again found himself flying, getting his wings with the Canadian Navy, and charged with the unlikely task of deterring enemy submarines from Lake Ontario. From there, however, he redirected his interest: instead of *making things fly*, he focused his energies on *preventing things from flying*.

Davenport's zeal for wind engineering coalesced when Carson Morrison, head of civil engineering at the University of Toronto, of-

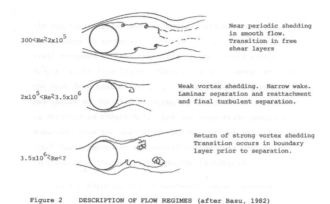

$300<Re\stackrel{<}{\sim}2x10^5$ — Near periodic shedding in smooth flow. Transition in free shear layers

$2x10^5<Re\stackrel{<}{\sim}3.5x10^6$ — Weak vortex shedding. Narrow wake. Laminar separation and reattachment and final turbulent separation.

$3.5x10^6\stackrel{<}{\sim}Re<?$ — Return of strong vortex shedding Transition occurs in boundary layer prior to separation.

Figure 2   DESCRIPTION OF FLOW REGIMES (after Basu, 1982)

Figure 16. An illustration of vortex shedding, or the formation of eddies in the wake of any bluff object situated in a steady flow of wind. Courtesy of the Boundary Layer Wind Tunnel Laboratory.

fered him a summer job. A television transmission tower was giving rise to some alarming vibrations, wobbling and swaying on a hill in Rimouski, a small town in Quebec overlooking the St. Lawrence River. The tower vibrated so violently it caused fatigue cracks in the foundation. It was also disturbing to onlookers and disrupting to Quebecers' cherished TV-viewing habits, causing ghosting, a doubling or tripling of images transmitted to the screen. Davenport went on a mission with his professor to investigate. "It was a puzzle why something purely circular should oscillate so violently," he recalled. When they arrived at the scene to conduct their investigations, the more immediate problem was that there was no wind. The duo thus contrived to mimic the wind and "excite" the structure themselves. Professor Morrison, being the weightier man, climbed to the top of the tower and lunged back and forth to induce vibrations, while Davenport measured the motion from below. The trouble, it turned out, was a dynamic form of excitation known as vortex shedding, or the formation of eddies in the wake of any structure subjected to a steady flow of wind. This was Davenport's first professional experience of the wind's destructive forces.

Ask engineers to explain how a slender structure moves in the wind—whether a vertical or a horizontal structure—and they are apt to invoke an analogy with a musical instrument, likening the phenomenon to the movement of a violin string when plucked or bowed. All linelike structures—towers and bridges, stacks, masts, power lines—will lean or bend under the influence of a steady wind. "There are two principal differences between the response of 'line' and 'point' structures to a natural wind and both are related to the effects of gusts," noted Davenport. "First, a point structure is likely to have only one mode of vibration, whereas a line structure may have many modes, each of which may be excited by the wind. Second, a point structure is only affected by the temporal velocity fluctuations of the wind at a point, while for a line structure the spatial variations of the wind velocity across the span are also important."

In addition to this static bending in response to a steady wind, structures have a dynamic response to the turbulent gustiness of the wind. Faced with such winds, built structures, like violin strings, have natural frequencies at which they vibrate. When incited by a quick gust of wind, a tower's response—like a string's when played—is not only to vibrate but also to resonate, reinforcing, prolonging, and amplifying the oscillation beyond the initial response. Another informative analogy is a child swinging on a swing. This is a simple "first mode vibration," going back and forth, back and forth, as a building or a bridge does in a turbulent gusty wind. To sustain the swing's desired motion and generate momentum, the child must pump or be pushed in sync with the swing's cycle. Just a little bit of input, a small tap at the top of each cycle to offset the energy lost by friction, makes the oscillation bigger and bigger and the swing climb higher and higher.

The wind acts much the same way on built structures. Imagine the wind's gustiness as a stormy sea with lots of choppy wavelengths, short and long. More likely than not, there will always be some wind wavelength hitting the building at just the right frequency to excite motion at the structure's natural frequency. This is called the "resonant wind frequency." If this particular wind wavelength rhythmically hits, say, the Rimouski TV tower in Quebec, the tower will sway alarmingly, traveling on an excursion even greater than that caused by steady wind. Gusting up against a building, like ocean waves against a breakwater, and hitting in sync with the structure's natural

Figure 17. Davenport with his wife, Sheila, in Bristol, 1959. Courtesy of Sheila Davenport.

frequency, the wind causes the structure's oscillations to grow and take on a life of their own.

The structural engineer's concern is to mitigate this motion. The Rimouski tower, Davenport and Morrison found, vibrated with a period of 7.5 seconds per cycle. The prescribed remedy was to stiffen the tower by installing guy wires, tensioned cables that work like ropes stabilizing a tent. This raised the tower's natural oscillating frequency, which in turn raised the wind speeds that would excite the structure to values that were not likely to occur, or at least not very often, given the local climate.

This little project in Rimouski proved seminal for Davenport. It inspired his master's thesis at the University of Toronto and later a paper often cited by colleagues, right down to the specifics of its series number—*Wind Loads on Structures*, Technical Paper No. 88 of the Division of Building Research of the National Research Council of Canada. It was published in 1960 following a stint at the NRC, where Davenport shared an office with two others, though he alone required an extra desk for all his books and papers on wind loads. The paper's preface states simply that wind forces on structures depend on two factors, on the velocity of the air (meteorological information) and on the shape of the structure itself (aerodynamic information). Grounded in an extensive survey of the existing literature on wind loading requirements in Canada's national building code, this epistle of sorts was assertive and full of recommendations. Davenport could see a long list of problems that needed solving.

In search of answers, he embarked on his doctorate in 1959 at the University of Bristol. This was a fortuitous locale, home to the Clifton Suspension Bridge, which became a much-loved loafing spot in high winds. The newly married Davenport was even successful in persuading his Canadian wife, Sheila Rand Smith, to accompany him there for informal investigations during storms. For the more formal research, Davenport's adviser and colleague was Sir Alfred Pugsley. Pugsley had been knighted for his leadership in structural and aeronautical engineering during World War II, and Davenport had selected Bristol chiefly because of Pugsley's reputation. He was a pioneer of research on safety. His ideas reflected a clarity and elegance of thought. He once wrote a treatise about wind in trees. And he was the author of a thin volume titled *The Theory of Suspension Bridges*. This latter solidified Davenport's own interest in the subject, an interest that quickly became an enthusiasm.

In particular, Davenport took note of Pugsley's work on the problem of aircraft flutter, a problem that increased as aircraft speeds accelerated. Prior to World War II, Pugsley had worked in the Airworthiness Department of the Royal Aircraft Establishment, where he conducted rigorous investigations into the dangers of flutter in aircraft structures. He studied how turbulence excited aircraft wings, causing them to resonate and twist. He derived new design criteria for wing stiffness and "aeroelasticity," a term he coined to describe the combination of aerodynamic effects and the elasticity of a structure. In the run-up to World War II, Pugsley was given the special responsibility of determining the aeroelastic properties necessary for new military planes, including the wing positioning and stiffness of the *Spitfire* and the *Hurricane*. Davenport, standing on Pugsley's shoulders, hypothesized that wind turbulence would also act on tall buildings and long bridges, and that there might be a parallel aeroelastic solution.

"One of the things that was very perplexing was how to deal with the ever-present problems due to turbulence," Davenport recalled. "Turbulence was an aspect of the wind which hadn't been dealt with in a very satisfactory way. Turbulence was muddled. It was chaotic. It fluctuated and you couldn't predict it. It was very indefinite, you didn't know how big it was, or what the frequencies were. This was turbulence. And how to summarize it neatly and compactly in such a way that engineers could use it, and design for this kind of flow, was not readily apparent. This constituted at least half the questions on my long list."

Pugsley gave Davenport free reign to explore. And so, in his PhD thesis, he sought to address a slew of questions on his list about turbulence that had hitherto gone unanswered:

> *What were the strengths and sizes of gusts and eddies in the*
>   *atmosphere?*
> *How did the nature and texture of the ground affect the wind?*
> *What was the wind like in cities?*
> *How did the wind vary with height above the ground?*
> *How did slender structures, tall towers, and long span bridges, pum-*
>   *meled by gusty winds, respond statically and dynamically?*
> *What were the risks of extreme winds, the chance of catastrophic*
>   *hurricane-force winds with return periods of several hundred*
>   *years?*

Davenport contends that much of his success was due to luck. He was in the right place at the right time. For starters, rudimentary aeronautic theories about turbulence had been developed in the 1930s. With the increasing availability of computers in the 1950s and 1960s, it became possible to compute statistical properties of turbulence. "Suddenly what burst in on the scene was a way to capture these chaotic qualities of turbulence into a rather neat and tidy statistical framework," he said. That was the essential thing—that there was a framework provided by aeronautics research for organizing the problems pertaining to turbulence. Davenport saw the way, as he recalled, "to apply these theories to more mundane structures, like chimneys and buildings and bridges. It was not so much inventing new knowledge as transferring knowledge just coming to light in another field and applying it to the civil engineering field." Nonetheless, he admitted with characteristic modesty, that it was "a very satisfactory step forward—that suddenly these terribly untidy problems became extremely tidy and manageable."

Anticipating his world-renowned boundary layer wind tunnel even in his Bristol days, Davenport attempted to bolster the statistical analyses with some crafty experimental technology. He commissioned the construction of a gust tunnel from a company specializing in the manufacture of church pews. His fellow grad students teased him about this contraption, and rightly so, because it never really worked. It was a basic wooden box, eight feet wide, with slats on either side that, when exposed to a jet of air, generated two independent flows amounting to a gust, if not a terribly realistic one. Davenport, nonplussed with the end product, soon gave up on his wooden wind tunnel. Still, he was onto something.

In 1961, continuing to refine his PhD thesis, honing it with a litany of big formulas pressed into the pages littering every room of his house, Davenport was offered a faculty position at the University of Western Ontario in London. The job brought the engineer and his wife back to Canada, along with their first son, Tom. Spending time in a bar on the ocean liner during the trip home, he wrote up his thesis, which contained seeds for much of his future work. The 237-page document demonstrated that the natural wind was turbulent, and that the mean velocity increased with height. These characteristics of the wind were well understood in meteorological circles of the day, yet were largely ignored in the investigation of wind forces on buildings. "Transference was the key to my success," he always said. He transferred knowledge from the fields of meteorology, structural engineering, and aeronautics to what would become known as the field of wind engineering.

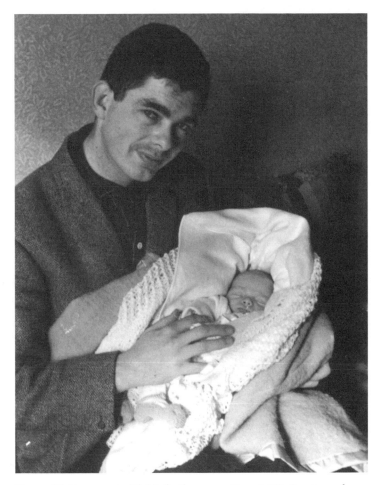

Figure 18. Davenport with his firstborn son, Tom, 1960. Courtesy of Sheila Davenport.

Davenport's success also hinged on simplicity of his professional style. His guiding philosophy was to keep things simple. His inattention to detail at times meant that hot-off-the-press handwritten lecture notes, and sometimes even published papers, skipped over seemingly important steps in his argument—often to the bafflement of students, researchers, and established engineers alike, who tried to retrace his path. (He was known as a gifted teacher on every level, even of practicing engineers in specialty courses for which he was in demand to teach internationally.) Details were lost, at the mercy of typos, since Davenport had no time or patience for proofreading. Nevertheless, in the big picture the answers and the insights were always right. He saw the forest without getting distracted by the trees (and he felt that leaving students to fill in the trees was good motivation; and to good effect, since at the end of a course he was known to proudly and drolly announce that his students had "Passed wind!").

Pulling out the essential information, not worrying about what was missing—that was quintessential Davenport. He was happy to capture 90 percent of a problem, sacrificing some specifics for the sake of simplicity. In this fashion, the initial formula he put down caused his mentor Pugsley some consternation—the elder statesman wondered what he'd unleashed on the world. Davenport's theory on the statistical analysis of wind forces on structures was impressively innovative and elegant, if a bit vague. His formulation rested on a naïve yet crucial guess about how winds accelerate as they go around the body of a structure. But then again, without this crucial approximation the phenomenon of wind turbulence would have remained mired in minutiae. And Davenport wasn't a fluid dynamicist; it fell to fluid dynamicists to improve on the fundamentals he set down. Sir Julian Hunt subsequently showed with his own method that the wind acceleration term in Davenport's formula was incorrect. Davenport's acolytes, however, are quick to come to his defense. "Alan's success was his ability to concentrate on primary issues, those that mattered, and pay less or no attention to secondary ones," says Nick Isyumov, one of Davenport's early PhD students, who made his entire career at the lab (where he is now a consulting director). "This I believe is a mark of a genius and not an attribute of a careless person or a bon vivant," says Isyumov. "The key to his success was his separation of the wind load into a mean or static component and a time-varying or dynamic component due to turbulence. The concerns about the stretching of the wind eddies, as the flow accelerated around the building, are a secondary consideration and one which has little influence on the end product." And in the end, the indefatigable Davenport, with the perfect combination of subtle diplomacy and relentless

Figure 19. Davenport amid a "Gotham City" of international buildings tested at the lab. Courtesy of the Boundary Layer Wind Tunnel Laboratory.

determination—proposing slews of new ideas and proselytizing their importance with inimitable panache—thereby started nothing short of a revolution in how civil engineering dealt with wind.

Seed by seed, Davenport's influence fostered the modern science of wind engineering. Boundary layer wind tunnels sprouted in Australia, Brazil, China, Denmark, France, Germany, Hong Kong, Italy, Japan, Korea, Norway, Thailand, Singapore, Spain, and Switzerland, as well as in the United Kingdom, the United States, and even at home in Canada. The leading practitioners all sought advice from Davenport, the internationalist and the inspiration. It is fitting, then, that for the logo of his lab Davenport chose the maple key, the tree's fruit that spins in the wind, floating great and small distances before reaching the ground and seeding another tree.

II
Tall and Taller Towers

On a typically wintry morning in January 1964, amid a mess of papers on a computerless desk, a black rotary telephone rang in Professor Davenport's basement office at the University of Western Ontario. Occupied at the moment with the chore of marking first-year students' assignments, Davenport answered the call that would make his career.

At the other end of the line Davenport found John Skilling, a partner in the Seattle-based structural engineering firm Worthington, Skilling, Helle & Robertson, and project engineer Les Robertson "singing a duet," as Davenport recalled, expressing their concerns about the possible effects of wind on plans for two tall buildings—both 110 stories tall, taller than the Empire State Building, and each double its volume. The project, Skilling cautioned, was highly confidential. He inquired as to whether Davenport, not yet thirty-two but already a pioneer in the field of wind engineering, would join their engineering team and run tests on these twin towers, which would become the tallest skyscrapers in the world, the World Trade Center. Davenport could hardly help but swoon. A modest man with not a bone of braggadocio in his body, he was astonished by this request. "Symbolically I fell off several chairs. It was the shock of my life."

Within days, Skilling and Robertson traveled to meet Davenport, not at the university but at Hotel London, again in the interest of confidentiality. Davenport impressed the men with his expertise. And for good measure, when Skilling discovered he'd lost a button from his suit jacket, Davenport obligingly ran to fetch a replacement. Skilling and Robertson left early the next day.

In short order, Davenport commenced two years of commuting to Colorado and New York, but first to Seattle. At Skilling's request he moved his family there for the summer of 1964, his wife Sheila five months' pregnant. They returned home just in time for the fall semester with nine-day-old baby Clare, their fourth child. When en route to New York, Davenport traveled by night train, waking in the sleeper car with a bedside view of the Hudson River. He arrived in time for early luncheon meetings at an oyster bar in Grand Central Station, usually with Robertson alone, who had relocated to open the firm's New York office (which subsequently became Leslie E. Robertson Associates). These were heady days, the highlight of Davenport's career. This seminal project established him as the father of modern wind engineering. The discipline in its modern form did not exist prior to the early 1960s, and the wind studies on the Twin Towers set a high benchmark.

Davenport had debuted on the international engineering stage only half a year earlier, in the summer of 1963. He attended a symposium, "Wind Effects on Buildings and Structures," at the National Physical Laboratory in Teddington, England, organized by Kit Scruton, who had made pioneering studies investigating the Tacoma disaster. The gathering drew three hundred attendees from twenty countries—the wind engineering giant was awakening, as Davenport recalled—though the conferees presented a mere twenty-four technical papers at this inaugural international conference on wind engineering (the conference is now held every four years). Davenport, however, had been allotted more than his share of time to present two papers, one on gust pressures and the other on the meteorology of the atmospheric boundary layer, both included in the conference proceedings published by Her Majesty's Stationery Office. His presentations were well received, he recalled, save for some incredulous questioning by someone from the Met Office. He carried on undeterred—Buddhist in temperament, he was a strong believer in keeping one's powder dry. Six foot one and standing pencil straight with a head of barely controllable dark brown curls, he was circumspect, measured in his actions and interactions. He earned a reputation for not saying a lot at conferences, and hence his colleagues listened all the closer when he did decide to speak.

During Davenport's Teddington lecture on the meteorology of the atmospheric boundary layer, he described his latest experimental work investigating the interplay of tall structures in strong winds, probing more unanswered questions: *What wind forces shear the base of a structure? What forces overturn a structure? What forces twist a structure? And to what extent do structures respond statically? Or dynamically?* Before the days of his wind tunnel, Davenport sought answers to these questions with fieldwork in the natural wind. He contrived a jerry-rigged experiment in an open expanse on university property. He mounted a 12-by-12-foot board, attached a load cell, and just windward of the board he set up an anemometer, an instrument that measures wind speed (derived from the Greek word for wind, *anemos*). His goal was to determine the correlation of wind forces with various levels of turbulence in the atmosphere.

Davenport lamented the complications that arose from conducting experiments outdoors. For one, while he thought this billboard for science might be less of an eyesore were it painted a cheery primrose yellow, the university's Buildings and Grounds Department disagreed, deciding that basic black would be more becoming. There was also Canada's inclement climate, given that the experiment launched just in time for winter. The tests, of course, had to be run whenever the wind was blowing hard, and thus cold. Nonetheless, the entire engineering machine shop sallied forth into the field enthusiastically. Experimental procedure entailed driving out to the field with all the recording apparatus, hauling the equipment across a snowy field on a toboggan, connecting a 500-foot cable to a generator for power, checking the board for icing, and then, finally, taking measurements. "Sometimes the wind blew from the correct angle, sometimes not," Davenport recounted. "Some useful results were obtained from these experiments, but progress was slow." It was during a snowstorm that he had his Eureka moment about building a wind tunnel.

Davenport's pluck and ingenuity grabbed the audience's attention at Teddington, particularly Les Robertson's. With the World Trade Center project already in the early stages of design, Robertson had

Figure 20. Davenport measuring the wind in an experiment on university property, 1961. Courtesy of Alan Noon.

done some historical research, looking at New York's landmark buildings. While the Empire State Building had been designed for a static horizontal wind pressure of 20 pounds per square foot, the World Trade Center was designed for at least double that. And at that, Robertson knew he was facing a steep learning curve. The Twin Towers were unlike anything ever built before. On the one hand, Robertson knew that the very presence of boundary layer winds had been proven to be the savior of high-rise buildings—without the benefit of broken and turbulent wind, as opposed to a static, steady wind, structures would have a tendency to self-destruct. On the other hand, Robertson was well aware that his towers would reach extraordinary heights where the wind's effects were as yet a complete unknown.

The architect of the World Trade Center, Minoru Yamasaki, was known as a master of "romanticized modern" and as an acrophobic, though his fear of heights did not stop him from building upward. He had recently designed the twenty-story IBM Building in Seattle with a façade that made it look as if it wore a pinstripe suit, a stylistic flourish also expressed in his design of the World Trade Center. When the World Trade Center project was announced to the public, mere weeks after Davenport received that fateful telephone call, Yamasaki eloquently stated his vision. "Paramount in importance is the relation of world trade to world peace," he stated, adding, "though the need to satisfy these spiritual and emotional requirements is preeminent, the scope of the World Trade Center demands a positive response to an increasingly important and practical challenge for architecture, that of advancing the art of building." The question of how to deal with the wind's effects on Yamasaki's advanced design had caused Robertson many sleepless nights. He went to Teddington in hopes of quelling his anxiety. "I went seeking genius, I went seek-

Figure 21. Hoisting of the American flag (center) and a Christmas tree (left) at the topping-out ceremony for the World Trade Center's North Tower, December 23, 1970. Courtesy of Leslie Robertson.

ing a guru," recalls Robertson. He was much relieved to find Davenport, an engineer thinking leaps and bounds ahead of the rest.

The advancing art of the modern skyscraper found an early advocate in the architecture critic Montgomery Schuyler. Writing on the evolution of the skyscraper in *Scribner's* in 1909, he defended this new structural form as a legitimate architectural expression of the times, but added a wind-related caveat: "The well-meant efforts to fix a limit by legislation to the altitudes which are converting the slits of street between them into Cimmerian and windswept ravines have thus far turned out to be either chimerical or futile and ridiculous." The biggest hurdle in designing ever-taller buildings was balancing their interplay with the wind, second only to the challenge of transporting lazy tenants or patrons who would never climb more than four flights of stairs (a problem solved when Elisha Graves Otis increased faith in the elevator by equipping it with a safety brake that could catch the lift in case the supporting cable snapped). When wind hits a structure in its path, it pushes the structure into motion. And as it makes its way around the obstacle, the wind goes one way, it goes another way, and in its wake there emerges a pressure differential that only causes the structure to move more. The trick is to devise a structural system that respects the wind, one that, metaphorically if not physically, rises above.

In the latter half of the nineteenth century, steel prices dropped, prompting a change in technology that allowed increases in height. Instead of the traditionally weighty and thus wind-resistant cast-iron frame around a structure's perimeter carrying the burden of the building's load, an internal gridlike skeleton of lighter, stronger steel beams and columns became the design of choice. Known as column-frame construction, this steel framework was implemented in many New York buildings during the skyscraper boom of the 1930s, including the Empire State Building. The advancing art of skyscraper design also converged with a booming economy and rising real estate prices that put a premium on space. As the architect Louis Sullivan, "father of the skyscraper," famously remarked, "Form ever follows function."

Soon enough, height hit a ceiling. While steel had allowed great vertical gains, its lightness also meant the structure lost its innate ballast, paradoxically making it weaker against the forces of the wind. Similarly, the transition from bolted to welded connections, from flexible to rigid joints, reduced a building's ability to dissipate the unwanted energy of motion, its capacity for "self-damping." Tall buildings had begun their wind dance.

The next iteration in skyscraper design, the next advance in height, came with the invention of the tube structure. At the heart of the design for the World Trade Center was a gigantic perimeter tube, cantilevered up from the structure's foundation, with an inner tube housing the elevator and services core. The design made economic sense since the absence of interior supporting columns increased the rental area, and structurally it provided a more unified defense against the lateral force of the wind. Using the tube structure, the World Trade Center towers would climb to record heights. However, the exceptional inherent structural strength of this innovation not-withstanding, Skilling and Robertson knew they had to proceed with caution, with a wind research program that went well beyond the standards of the day.

For the first half of the twentieth century, wind speed concepts in the United States were based primarily on the notion of the "fastest mile of wind." Prior to the availability of electronic or digital recorders, the fastest mile was measured using a seemingly whimsical device, but in fact state-of-the-art technology for the time: a rotating cup anemometer that made a mark on a steadily rolling chart each time the cup turned through an actual mile of wind. If the wind speed was 60 miles per hour, then one mile of wind would take one minute to pass by. By measuring the shortest distance between marks, the investigator could estimate the highest wind speed, and this would be used as the basis for the "design wind speed," the specified wind speed that a structure should be built to withstand.

Measurements also came to consider not just the fastest winds but also their recurrence intervals, their "return periods"—that is, how often the high winds were likely to occur, once in fifty years or once in five hundred. Though broadly speaking, when the Twin Towers were under design, analysis criteria remained rather fuzzy. In selecting a design wind speed, engineers would typically consult the New York City building codes, which were based on a steady wind concept, as if the wind pushed with a constant, sustained, static force against a building (the Empire State Building's was 20 pounds per square foot, pertaining only to 100 feet and above; if following those standard codes, the World Trade Center's would be 45 pounds per square foot). Typically there would be no accounting for the dynamic and turbulent aspect of the wind or for its inherently stochastic nature, its random, nondeterministic behavior, blowing at one speed and direction one day and at an altogether different speed and direction the next day, let alone hour to hour or minute to minute. The temperamental wind at any moment in time provided no predictive power for what might come next. Thus, in Davenport's view, the New York City building code requirements weren't the least bit credible, though they weren't any worse than standards in other cities.

Davenport's simple, elegant, and unprecedented wind engineering innovation was to model the stochastic wind—investigating what wind forces would shear the base of the structure, what forces would twist the structure, and what forces would overturn the structure—and to model it in two distinct yet complementary ways.

First, he would model the wind's interaction with the structure. This would be difficult to model mathematically, so he proposed modeling it physically in a wind tunnel, where the physics of both the boundary layer wind and the structure could be accurately reproduced. The variables affecting dynamic excitation were myriad, such as the turbulence of the wind, bumping up and down around and about a structure, and the fluctuating winds that develop in the wake of the structure, like the wake of a boat adding to the choppiness of a lake's waves. For this type of modeling—modeling on the micrometeorological scale, replicating gustiness from second to second or minute to minute—Davenport took into account the nittygritty details hitherto not given much consideration, details such as local topology, terrain roughness, wind directionality, as well as the geometry of the structure, the wind flow around its perimeter and that of its neighbors. Second, Davenport would create a computerized rendering of regional wind climates—weather fluctuations on a mesometeorological scale that vary over the course of years and months, days and hours—to develop a statistical model of the real wind speeds and directions likely to occur at a site. For this type of modeling he welcomed all the data he could get his hands on, from existing anemometer records, to upper-level winds measured by his own balloon-piloted rawinsondes, to data on storms generated by computer simulations.

This meticulous approach would provide a representation of wind speed more nuanced and more accurate than the fastest-mile metric.

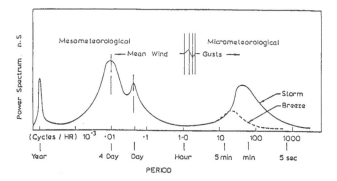

Figure 22. Davenport's graph adapting the Van der Hoven spectrum of winds, ranging from mesometeorological winds, the year-over-year climatic patterns such as seasonal monsoons, to micrometeorological winds, such as gusts and breezes that vary from moment to moment. Courtesy of the Boundary Layer Wind Tunnel Laboratory.

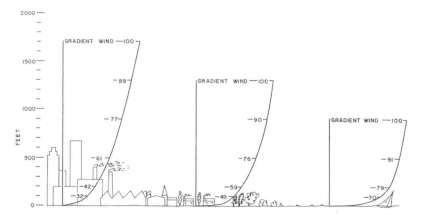

Figure 23. Davenport's illustration showing the mean velocity profile of wind over various terrains—city, suburban, open country, and open water. The presence of greater terrain roughness, in the city for instance, causes the height of the boundary layer to increase, meaning high-speed winds occur at higher altitude. Courtesy of the Boundary Layer Wind Tunnel Laboratory.

Wind speed would be comprised of two components: a mean value, the average speed over a specified time period, say ten minutes or one hour, corresponding to the wind speed that could be set in the wind tunnel experiment; and a turbulent value, the fluctuations in time and space around the mean, which would occur naturally in the wind tunnel simulation if the terrain were properly modeled. These variations would be captured with "velocity profiles," composite measurements tracking the mean wind speed from zero, at the ground, to its maximum at 1,500 to 3,000 feet above the ground, varying according to the roughness of the terrain. And associated with these mean velocity profiles would be profiles of turbulence intensity, turbulence being greater near the ground and over rough terrains, and decaying to nothing at higher altitudes.

By wedding these wind tunnel simulations for all possible wind directions with historical statistics on hourly wind speeds, Davenport could cut to the core of reams of data and predict how structures would behave in the wind. His wind tunnel, in essence, would be an analogue computer, programmed through the climate model to clearly answer complex questions for the design engineers about turbulent wind response—questions such as how often design loads or their associated pressures, deflections, and accelerations would occur. "The engineering art," Davenport once observed, "is to ensure that the qualities important in the design are reproduced reasonably accurately. That is the stock and trade of boundary layer wind tunnels."

But to put all this into action, Davenport needed a wind tunnel. And no ordinary wind tunnel would suffice. In a standard aeronautical wind tunnel, the object of study, usually an airplane's wing or engine, is mounted on a pedestal-like platform. The model is positioned in the middle of smooth uniform flow to simulate the silky "laminar" flow that an aircraft usually meets while flying through clear skies at high altitudes, above 5,000 feet, where air is relatively smooth. This is how the Empire State Building was tested, at the National Advisory Committee on Aeronautics, the forerunner to NASA. A smooth ride, however, is not representative of the wind at lower altitudes, at the lowest slice of the atmosphere, the planetary boundary layer. In these parts, air currents are rough and erratic as they come in contact with the thermal and physical obstacle course that is the surface of the earth—air currents stop and start, speed up and slow down, swerve and turn, and the usual horizontal flow is tripped up, sent flying into three-dimensional gusts.

A Danish engineer by the name of Martin Jensen was the expert on this subject. Having studied wind and plant growth, as well as the aerodynamics of locust flight, Jensen also wrote a more apropos paper that caught Davenport's eye, titled "The Model Law for Phenomena in the Natural Wind." Therein he identified for the first time the importance of the boundary layer and presented a law for modeling it (introducing what is now known as the Jensen number). "A great many technical circumstances depend on the wind in nature and cannot, or only with great difficulty, be analyzed except through model tests," Jensen wrote, continuing to remark on the misleading investigations to date that had neglected to incorporate turbulent flow. "The correct model test for phenomena in the wind must be carried out in a turbulent boundary layer, and the model-law requires that this boundary layer be to scale as regards the velocity profile." In Jensen, Davenport found not only a mentor and a lifelong friend but also a viable strategy for conducting realistic wind tunnel testing.

Practically speaking, three options presented themselves for a wind tunnel facility. There was the wind tunnel at the University of Washington that had been used in the Tacoma post-mortem. There was Kit Scruton's wind tunnel at the NPL in Teddington. And Davenport had recently read in a meteorological journal about a bona fide boundary layer wind tunnel facility recently constructed as part of the Fluid Dynamics Laboratory of engineer Jack Cermak at Colorado State University in Fort Collins. Cermak's interests were slightly different from Davenport's. The Colorado wind tunnel, financed by the U.S. Army, had been built to study the fluid dynamics of boundary layers—the dispersion of gases and air pollutants, the diffusion of radiation clouds, the aftereffects of chemical weaponry. Nevertheless, Cermak was curious about Davenport's intentions and agreed to lease his wind tunnel for the World Trade Center's experimental program. The facility would need some renovations—the roof had to be raised to decrease the pressure gradients—but this monster of a scientific instrument

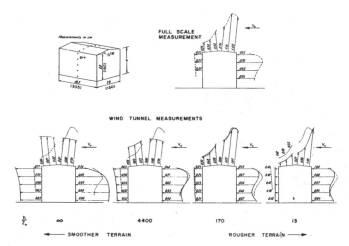

Figure 24. Results showing early boundary layer wind tunnel investigations in the 1950s by the Danish engineer Martin Jensen. Based at the Technical University in Denmark, Jensen built a boundary layer wind tunnel in which he tested a model of one small building—about the size of a large doghouse—and compared the results with the results of full-scale tests. The top graphics show pressure measurements from full-scale measurements, while the lower figures show pressure distributions measured in the wind tunnel, ranging from smooth terrain over water (left) to increasingly rougher terrain such as found in a city (right). The pressures measured in the wind tunnel are shown with arrows, while the full-scale measurements from the top figure are superimposed on each of the lower figures. This reveals how the best agreement is found when model terrain matches the full-scale terrain in terms of the relative heights of the roughness and the building (graphic second from right at bottom), and hence indicates that testing in smooth flow does not accurately reflect the reality of full-scale measurements. Courtesy of the Boundary Layer Wind Tunnel Laboratory.

would more than fit the bill. Fed by a mangle of tubes and electronics, the rectangular contraption hovered above the floor on a procession of stilts. A four-blade propeller, courtesy of a World War II–vintage P-38 fighter plane, cranked out the modeled wind at a maximum of 35 miles per hour, blowing northward along the length of the chamber for 100 feet, then through an interregnum of coils and baffles to calibrate its temperature, around two corners, and then back southward toward the test section, narrowing to six feet by six feet in dimension. Plexiglas lined one side of the tunnel, allowing a view of the proceedings from a control room full of dials and knobs and wheels, and a red button that turned all the machinations on.

Figure 25. Engineer Jack Cermak (right) with Davenport and aeroelastic models during the World Trade Center wind studies at Colorado State University. Aeroelastic models, measuring the intertwined forces of inertia, elasticity, and aerodynamics, had never been applied to the design of buildings before their use during the WTC tests. Situated here without any proximity models replicating surrounding terrain, these tower models were used to investigate the interaction of the two structures, specifically the effects that each tower's wake had on the other tower. Following Davenport's lead in using a wind tunnel to study the built environment, Cermak coined the term "wind engineering" in the 1970s and provided a formal definition: "Wind engineering is the rational treatment of interactions between wind and man and his engineered works on the earth's surface." Until this time the field was known as industrial aerodynamics. Courtesy of Cermak, Peterka, Petersen Inc.

Davenport now needed grist for his mill. His deceptively simple idea was to replicate a degree of realism more detailed and more precise than hitherto available, to model both the structure and the wind based on data drawn from the real world. The model of the World Trade Center buildings would replicate their mass, stiffness, flexibility, and natural frequency. The wind tunnel would replicate the speed, force, direction, and capriciousness of the wind. It sounded simple enough. But amassing the data and executing the modeling would be an intricate and time-consuming endeavor.

Davenport reproduced the Twin Towers and their environs at a scale of 1:500—greater detail than that of a military general mounting a sandbox mock-up of battle. He procured site plans for a 2,000-foot radius, as well as maps of the city's urban landscape, and fashioned a replica of the terrain with minimalistic materials. Rough carpeting stood in for the Upper Bay and the Hudson River, the open water fetch—fetch being the geographic term for a length of terrain over which wind has traveled. The carpeting extended almost the full length of the wind tunnel's 100-foot test section. Randomly distributed blocks of wood or Styrofoam, ranging in height from one to three inches, mimicked the Jersey City fetch. A similar patch of slightly taller blocks represented Queens and Brooklyn, while Manhattan required blocks taller still.

This Manhattan zone, being in the immediate vicinity of the Twin Towers and within the targeted test area, sat on a four-foot turntable so it could be rotated and subjected to wind from all angles of attack around the compass. Factoring in the wind's directionality was a novel concept, and a crucial one; on many occasions since it has been the deciding factor in whether a project is built or shelved. The common practice at the time was for meteorologists to take the set of highest recorded wind speeds for the area and then predict a "once in fifty years" wind speed. This worst-case scenario wind would be thrown at the structure from all directions in an aerodynamic wind tunnel, and then the worst-case results dictated the design. By contrast, Davenport's method, which used probability theory, would measure data for each wind direction, factoring in each direction's relative importance (see the sidebar, "Davenport's Probabilistic Prediction Method," page 58). With the turntable sliced into twenty-four or thirty-six equal segments, a rotating drive stepped through all wind directions, one total revolution requiring up to fifteen to twenty hours of testing.

Finally, there were the all-important models of the Twin Towers themselves, the bull's-eye on the turntable target. Davenport, a dapper sartorialist when the occasion warranted, fashioned replicas tailored to Yamasaki's trademark pinstripe piping. He made two sets of models, a rigid model and an aeroelastic model. The rigid model, as its name suggests, stood stiff. Rigged with a multitude of pressure taps, a strategic polka-dotting of holes on the exterior, it would ab-

Figure 26. World Trade Center pressure models surrounded by a proxy model of Manhattan. Pressure models measuring both pressures and suctions on the façade were standard procedure of the day, but traditional testing used manometers, capturing only mean pressures and suctions. By employing pressure transducers for the Twin Towers wind tests, Davenport measured both steady-state and dynamic pressures. Courtesy of Leslie Robertson.

sorb the wind's force and measure the impact, the weight of the wind pushing on the façades. Any change in air pressure at a tap mouth created a chain reaction: the pressure change would travel through thin tubes leading down inside the model, below the turntable and wind tunnel floor, to mini-diaphragms equipped with strain gauges at the model's base. When pressure, force per unit area, deformed the strain gauge's thin foil sensor, the sensor would send electrical pulses from the gauge to a computerized data center. In seeking yet more

Figure 27A (facing page top). Contour drawing of mean external pressures based on the wind tunnel tests, indicating pressure measurements obtained using the model shown in figure 26. The drawing shows (at left) the positive pressures measured over the windward surface of the tower and (middle and right) the suctions created along the sides and the leeward surfaces of the tower. Courtesy of Leslie Robertson.
Figure 27B (facing page bottom). Vertical distribution of the pressures and suctions as shown in figure 27A. Courtesy of Leslie Robertson.

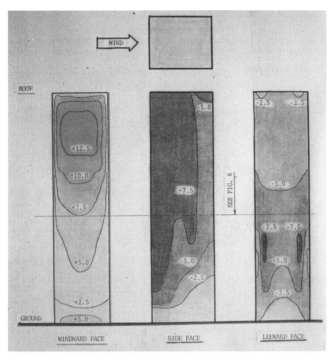

WINDWARD FACE   SIDE FACE   LEEWARD FACE

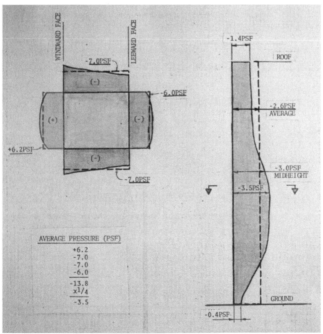

complex information, Davenport also created the aeroelastic models—aeroelasticity being a property reflecting the intertwined forces of inertia, elasticity, and aerodynamics. Models based on these concepts were standard in aeronautical testing and had been used to test the Tacoma Narrows replacement bridge in 1950, as well as the Severn Road Bridge in 1966. But an aeroelastic model had never before been applied to the design of buildings. Davenport's goal, as with the Rimouski tower, was to simulate the structures' dynamic characteristics, to project how the Twin Towers might sway in the wind.

To create the aeroelastic model, Davenport worked backward with some reverse engineering. The design engineers had predicted the mass of the full-scale building, its stiffness, natural vibrating frequency, moment of inertia, and rotational mass, its damping ratio, and its design wind speed. By mathematical sleight of hand—that is, by applying scaling laws—Davenport and his team then crafted models that were dynamically equivalent. The heart of the aeroelastic models resided in the base, hinged and mounted on gimbals—a pivoting support facilitating rotary motion, like a universal joint as found in a suspended compass. When exposed to wind, the model thus had the freedom to move and sway, to teeter and deflect about the two directions of the standard x-y-axes (this simple approach ignored torsion, considered unimportant for the slender towers, and approximated their first bending mode shapes as simple straight lines—fortunately, a very good approximation for many tall buildings). To replicate the towers' natural frequency, the engineers fine-tuned sets of springs beneath the base: the stiffer the springs, the less the model could move, and the higher the frequency (or the lower the period) of the vibration. And finally, electromagnets in

the base simulated damping: the stronger the electromagnet, the stronger the damping, and the quicker the model's sway died away. When subjected to the wind, strain gauges again measured the model's deflection, and the measurements went to the computer. These statistics on the model's dynamic response to the wind were then translated back through the scaling laws and applied to the building itself. Whereas the rigid model allowed a controlled dry run, the aeroelastic model was as close as an engineer could ever come to a full-scale dress rehearsal.

For the most part, this pathbreaking wind tunnel study had remained hush-hush, confidential, generating top-secret information of the utmost sensitivity. Yet the Fort Collins *Coloradoan* got wind of it, reporting that "twentieth-century Gullivers, in the guise of engineering scientists, are working in the large wind tunnel of Colorado State University's Fluid Dynamics and Diffusion Laboratory. Where else can one step over the Empire State Building and the even taller gleaming white towers of the future World Trade Center of New York? . . . [The technicians] step carefully from Broadway to the East River and beyond to the Atlantic—not in seven-league boots, but in stocking feet, mandatory footgear in the tunnel." Thankfully, the *New York Times* did not pick up the story—thankfully because public debates raged over the advisability of building these tallest of towers. And initial findings from the wind tunnel experiments might have appeared troubling to the untrained eye.

Initially, Davenport tested models in smooth laminar flow. He sat them near the propeller, and there they faced extremely high winds, equivalent to winds of 140 or 180 miles per hour at full-scale speeds. These would be the ultraconservative wind speeds that designers usually threw at structures, the worst-case scenario winds. Davenport included them because he wanted to run through the gamut of wind speeds, he wanted to take into account even the worst winds— providing a benchmark, of sorts, to the testing methods of yore. And so it was hardly surprising that these winds produced a hyperactive response. Davenport had commissioned an independent backup study at the Teddington lab, and there the models, designed by Scruton but subjected to Davenport's experimental protocol, came close to self-destructing. Even at the Colorado lab, the models painstakingly crafted by Davenport displayed discouragingly huge deflections, wild excursions, wobbling far afield from what would be deemed an acceptable path of movement in the wind. "We were cer-

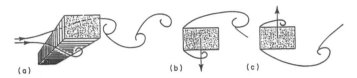

Figure 28. Davenport's illustration showing the effect of vortex shedding on tall buildings. When eddies are shed from a structure at nearly the same frequency as the structure's natural frequency, vortex shedding shakes the structure back and forth (indicated by the transverse arrows). Image (a) shows the flow over the top of the building, while (b) and (c) show the flow around the building—and it is (b) and (c) that generate the "shaking" of vortex shedding, also known as a "Kármán vortex street," after Theodore von Kármán, a Hungarian-born American aerospace engineer and physicist. Courtesy of the Boundary Layer Wind Tunnel Laboratory.

tainly getting huge amplitudes," recalled Davenport. "Just colossal." In these high wind speeds, the models displayed a "very, very peaked response." The models were enjoying an aeroelastic instability phenomenon whereby a structure's anatomy actually incites its own vibration, exacerbating rather than mitigating motion. Technically speaking, this translated into troublesome magnitudes of vortex shedding. The wind flow separated from one side of the building and rolled up into a little vortex, a miniature tornado tearing away as it traveled downstream. Meanwhile, the wind was making its way into the same formation on the other side of the building. And so on and so forth, from one side to another, pushing and shoving the structures this way and that like a gang of bullies.

As Cermak watched the models vibrating together like a gigantic tuning fork, he figured they were moving as much as an inch. At a scale of 1:500, that implied the buildings might move as much as 500 inches, or more than 40 feet. Luckily, the scaling laws required some corrections; the models represented only two degrees of freedom, so they didn't provide a perfectly direct correlation. But the magnitude of movement would be considerable. "Pick a number," Davenport quipped—which was to suggest that in a sense it hardly mattered. He and Robertson were just mucking about at that stage. With no precedent to speak of, nothing to stand on, they were finding their feet, testing indulgently, looking at every possible variable. Robertson would say, "What about this?"—what about those worst-case scenario winds?—but then Davenport would counter with "What about that?"

The key "what about" factor in Davenport's innovative modus operandi was to find out all he could about the boundary layer winds, the turbulent winds at the site, at the record-breaking height of the towers. He needed to model the natural wind realistically. "We knew nothing, really, about what wind strength we should consider for towers of this height," he recalled. "So now the spotlight [went] onto what the real wind speeds were at that height. This was the area in which we'd been rather ultra-conservative. We'd chosen a wind speed that was unusually high. Because we had to extrapolate from surface-based operations to the wind speeds at a height of 1,500 feet." The other crucial factor that came into play was "the question of how should you deal with wind direction," he said, "because to assume that the wind always comes from the worst direction was ultra-conservative as well."

Setting out on his quest to quantify the wind, Davenport compiled a database of surface wind speeds from all major weather stations on the East Coast from Maine to New York City. He collected climate data from Brookhaven National Laboratory on Long Island, where an archive retained records from a number of hurricanes. He obtained complementary information from hurricane hunters who flew through storms. He procured long-term observations from John F. Kennedy and LaGuardia airports, observations made by a rawinsonde, an instrument sent up in weather balloons to measure wind speed and direction. And—just his luck—the rawinsonde samples had been obtained near Manhattan and extended beyond the full height range of the towers.

Wanting more data still, Davenport launched his own fact-finding mission, sending up his own fleet of weather balloons directly from the World Trade Center site. He also deployed rapid-response anemometers, the Mercedes of anemometers, manufactured by Bendix-Friez Aerovane and equipped with autographic pen recorders. These were set up well above the rooftops of neighboring buildings, such as the 581-foot New York Telephone Company building and the Bank of Manhattan Trust at 40 Wall Street (now the Trump Building), with its gothic spire topping out at 927 feet. This was the first time the turbulence of wind above a city had been measured with any reliability. By comparing the characteristics of the various wind samples, Davenport could determine the effect of the city's built environment on the wind. And then he reenacted the relevant scenarios in the wind tunnel.

"It's not exactly Christopher Columbus," Robertson noted, looking back, "but we were moving into entirely new territory and studies like these were vital." For his part, Robertson rented a top-floor apartment in a building near the Empire State Building at 37th Street solely for its bird's-eye view of the boundary layer. Late at night, with the city's lights refracting in the fog, Robertson watched for airflow patterns that might give him an edge on the wind. He took some comfort in the fact that the wind flow he witnessed matched what he saw in the wind tunnel when a probe injected a stream of smoke into the test section, revealing the whirls of the vortices and the jumbles of turbulence for the naked eye to see.

With the wind data in hand, and advancing along the what-about-this, what-about-that spectrum of every and all possibilities, Davenport next tested the rigid and aeroelastic models in the meticulously sculpted turbulent flow, matching the design speeds, and from all angles of attack. As expected, at high wind speeds, the turbulence had the beneficial effect of suppressing the vortex shedding, breaking up the vortices and reducing the deflections, reining in the model's wandering excursions. But there was at once a counterintuitive effect of the turbulence. "If we added turbulence to the flow," Davenport said, "what we would find is that the vortex shedding was substantially reduced. The more turbulence we added, the more the response was reduced. But it also generated the much larger responses at wind speeds below the speed at which vortex shedding would develop. So this is a sort of a paradoxical situation that the lower wind speeds were, relatively speaking, producing higher dynamic responses than the vortex shedding at higher wind speeds." This was a buffeting response. Persistently buffeting the models, thumping them with gusts of wind, created a wider range of low wind speeds that excited the towers. Overall, the turbulence flattened out the very peaky response,

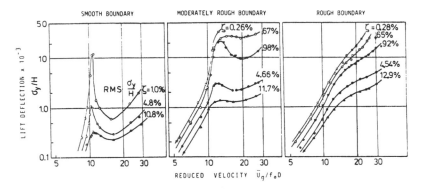

Figure 29. Davenport's graphs demonstrating that turbulence has the beneficial effect of leveling off a structure's peak deflections. Courtesy of the Boundary Layer Wind Tunnel Laboratory.

as Davenport described it, but it broadened the peak, widening the range of frequencies that excited the towers, producing a flatter and wider curve of response. The turbulent flow would create more motion at everyday wind speeds, causing more everyday motion, and though otherwise harmless, this everyday motion might cause motion sickness problems among the towers' occupants. There were now a few new problems that needed solving, with more tests in the wind tunnel.

ധ   ധ   ധ

By 1965, Robertson had relocated command central to New York. Davenport, back in London, commuted Monday to Friday. Aware that stress levels were climbing, Robertson at one point suggested (somewhat in jest) that they take in a disco. Davenport confessed he had never been. So they found a club and hit the dance floor. Davenport slowly started moving, gathering force, easing loose his inhibitions until eventually he waltzed around with abandon.

Despite the attempt at finding an occasional outlet, the stress-inducing preoccupations remained—to improve the towers' waltzing abandon, to improve their dynamic performance. To that end, Robertson had yet another what-about-this query he wanted to chase down. As if all the elaborate wind tunnel tests weren't due diligence enough, Robertson—known as a particularly gifted engineer, with a stellar reputation for his penetrating critiques and preparedness in assessing scientific data—masterminded another unprecedented precautionary step. He wanted to actually test creature comfort, the effect of the towers' sway on occupants. He wanted to quantify human tolerance for movement in a skyscraper.

As far as could be ascertained, any testing to date on motion sickness had focused on the high-frequency vibration of machinery or space capsules, or on the motion of ships and automobiles, but never on a skyscraper's notorious wind dance. Generally speaking, human perception of horizontal motion was known to depend on fluid movement in the inner ear and the resulting internal pressure changes. The intensity of the pressures depended on the jerk, the acceleration's rate of change. When the vestibular system, the balance system in the inner ear, sensed motion without the visual perception system receiving corresponding sensory data, these conflicting brain inputs would cause synaptic confusion, or motion sickness—nausea, dizziness, fatigue. But what was the exact point at which discomfort set in? What were the psychophysical effects of the buildings' motions on the buildings' inhabitants?

In August 1965, with the Port Authority's green light—and a generous add-on to the ever-expanding budget—Robertson's study of human perception thresholds of horizontal motion commenced, with the usual attention to confidentiality. Secreted away in an office building across from the courthouse in Eugene, Oregon, the experimental setup was rather creative. Ads in the local newspaper had promised complimentary checkups at the new Oregon Research Institute Vision Research Center. Takers walked into a stereotypical waiting room, where they were greeted by a receptionist and shown into an examination room—actually, one of two motion simulators designed by Robertson, modeling the average workaday office with a desk, a fan, and a door, but no window, since a window would have provided a visual cue, a way for the brain to orient itself with respect to the motion. Beneath the façade, the contraption was mounted on

Figure 30. The Kafkaesque control room of the motion simulator designed by Les Robertson to study human perception thresholds of horizontal motion. Courtesy of Leslie Robertson.

a wheeled platform driven by hydraulic actuators, which facilitated back-and-forth motion simultaneously in two directions on the horizontal plane. Inside, with a legitimate optometrist in cahoots, Dr. Paul Eskildsen began the test by instructing his subject to stand by a mark on the floor. "I am going to project some triangles on the wall," he told them, "and I would like you to estimate the height." As the subjects proceeded with the task, the room started moving. Robertson wanted to know how much movement it took to make people disoriented, dizzy, and ill. Scientifically speaking, human reaction to vibration had been classified into six comfort zones: 0—imperceptible, 1—faintly perceptible, 2—distinctly perceptible (not objectionable), 3—strongly perceptible (objectionable), 4—disagreeable (if prolonged, could be harmful), and 5—painful (definitely harmful even for short periods). Robertson was interested in what happened between zones 0 and 1, between imperceptible and just perceptible vibration. The examination room's oscillations each lasted 5, 10, or 15 seconds. At a certain point the tests were interrupted—if, that is, the subject hadn't already "popped," as one did in exclaiming, "Whoa boy! The room is wobbling." Dr. Eskildsen then inquired as to how his patients were feeling:

> "Goofy. I was kind of reeling around. It's a feeling of not being able to control my standing. Are you hypnotizing me? Because that's kind of sneaky."
> "I feel like I'm on a boat."
> "I think you're taking away my gravity or something."
> "It's unpleasant. . . . Maybe I'm on Candid Camera."

Over two weeks, 112 subjects endured rocking and tipping and sensations of having "rounded feet." By the end of each day, Dr. Eskildsen himself suffered from motion sickness. Yamasaki, Helle, and Robertson came to visit and took a ride in the simulator, driving it to maximum deflection. And Robertson erected a second simulator in New York, exclusively for the edification of Port Authority officials: he took a packing crate, decorated the interior to resemble an office, suspended it by four cables from a ventilation tower in the Lincoln Tunnel, and pushed it into motion manually, guided at the base by bicycle wheels. In both test settings, sensitivity varied dramatically from person to person, influenced by a number of factors such as body orientation, body movement, body posture, expectancy of sway, and the period and magnitude of oscillation. Ten percent of subjects noticed as little as two to four inches of sway. The average person sensed about five inches. The bottom line: the shorter a building's oscillations, the better. The less the Twin Towers swayed, the happier their inhabitants would be.

Fascinated by these studies, Davenport kept an eye on the proceedings but remained at arm's length. Meanwhile, at his behest the wind tunnel tests continued, in the quest to improve the towers' performance. He had to rein in the motion of the model towers. He had to ensure that the modeling—both of the structures, and also of the wind—gave an accurate representation of reality. If the aeroelastic model was too flexible, if the springs were too loose, or if the wind speeds were too conservative, these misaligned variables would misconstrue the towers' response. For a time, refining the modeling, especially of the towers, kept Davenport busy. "I was very insistent on getting the dynamics right," he recalled. "Designing dynamic models was my day and night pursuit." And, with all the wind data and his bespoke computer simulations, he also had to decide on a reasonable expectation of wind speed. "Eventually we reached the point that we had what we thought was a very rational description of the progression of [the model's] response as the wind speed rose," he said.

Once a logical modeling relationship had been established to everyone's satisfaction, attention turned to addressing the implications for the real-life towers. With the data generated from the models, Davenport and Robertson's design team derived the full-scale structures' performance in an iterative back-and-forth, juggling a bevy of variables—wind speed, for one, and turbulence and vortex shedding, as well as the cross-wind, along-wind, and torsional responses, with particular attention to the aerodynamic interference of the upstream tower on the tower downstream. Then, finally, they came up with the criteria used in the design, resulting in a few tweaks to the towers to get their best performance.

The wind tunnel tests continued for two years. The lab issued the final report (the last of three, and a doorstopper) in June 1966, a volume stuffed full with foldout charts and graphs and photographs, including a panoramic view of Manhattan taken from the site of the World Trade Center at the height of the tower tops, as well as FORTRAN statements from the computer, interspersed with narrative analysis. From Robertson's point of view, the reports didn't hand down any ironclad conclusions. Rather, they chronicled the painstaking what-about-this, what-about-that iterative process as Davenport and Robertson zeroed in on the best performance possible for the towers. The first report had stated the obvious, what Davenport and Robertson well knew: "[The] turbulent or 'boundary' layer has important consequences in the design of the towers of the World Trade Center." And over the duration of all the investigations, one thing became clear: if built as first designed, without the costly addition of extra steel to provide more strength and stiffness, the towers would be overly excited by the wind. These were optimized structures—optimized in that they would be constructed from the most effective minimum of concrete and steel—and they were incredibly lightweight. As a result, they possessed low self-damping. There was little inherent heft, little intrinsic dissipation of energy to steady them once they got moving. All in all, an alarmist might interpret this to be a huge problem, but it was really only a problem in the sense that the easy solution—to strengthen and stiffen the structures by throwing more steel at the design, increasing its structural damping and raising its resonant frequency—would be a huge expense.

Instead, the iterative back-and-forth continued, as Davenport and Robertson summoned all the data and investigated more palatable tweaks to improve the towers' performance. Here Davenport deployed what became known as his probabilistic prediction method, his theory that juggled all the interdependent variables statistically, using probability theory to calculate the odds that the structures would enjoy wind speed X, from direction Y (this was a pencil-and-paper method; it lent itself to computer analysis to speed up the process, but Davenport did much of the work by hand).

"It is evident," the final report noted, "that the response of the towers is highly sensitive to wind direction." If Davenport, in assessing the data, could convincingly reduce the odds, the likelihood, that the most deleterious wind speeds and wind directions would occur, let alone coincide, if he could shrink the hot zone, then he and Robertson could more efficiently equip the structures to face these critical winds—by, of course, testing again in the wind tunnel, tweaking again the design plans, testing and tweaking, testing and tweaking.

Davenport's Probabilistic Prediction Method

Turbulent wind is complex, chaotic, nondeterministic—in short, a stochastic system that is difficult to predict. One of Davenport's innovations was his probabilistic prediction method, investigating and analyzing the wind with the tools of probability, the mathematics of random phenomena.

As Davenport noted in his final report on the World Trade Center, "Because of the directional characteristics of wind pressure the probability distribution of wind pressure at a point on the exterior surface of the building is only indirectly related to the probability distribution of velocity. A specific value of wind pressure can arise either from a low wind velocity coming from a critical wind direction or a higher wind velocity coming from a less critical direction."

The method he devised for estimating the combined probability is as follows:

First, in the wind tunnel, measurements are made of the fluctuating responses from a model of a building subjected to turbulent boundary layer flow that reflects the correctly modeled wind from each direction around the compass. Using these data, the structure's response is mapped versus wind direction and speed.

Consider, for example, the pressure at one point on a building (see figure 31A, top middle image, with the dot indicating a pressure tap on a narrow rectangular building). A pressure coefficient—a dimensionless number that describes relative pressures in a fluid flow field—is measured for a complete range of wind angles. For each angle, a time history of pressures, which fluctuate even more than the wind, is generated (see figure 31A, top right). Over the equivalent of an hour in full scale (the scaled time is typically 30 seconds in the model testing), the largest positive and negative coefficients are chosen. Negative pressures, or suctions, are usually the most interesting analytically because they are the biggest.

For a single point on a building, the peak negative coefficients are then plotted in polar form showing all the wind angles, though pressure coefficients of large magnitudes are usually found only over a narrow range of wind angles (see figure 31B, the solid line that peaks out to the right in the polar plot). These coefficients can be carried over directly to full scale (since the pressure coefficient = pressure divided by 0.5 times the air density times the speed squared). And thus this polar plot can be used to answer the simple yet crucial question at each wind angle around the compass: What speed is required to obtain any specific pressure, such as 30 pounds per square foot? To answer the question, simply invert the equation to solve for the wind speed, knowing the pressure coefficient (speed = the square root of 2 times the chosen pressure level of 30 psf, all divided by the air density and the pressure coefficient at this angle). Done at every wind angle measured, this calculation allows a speed contour to be defined for this specific response level of 30 psf (see figure 31B, the dotted line in the polar plot). When the pressure coefficient is large, not as much wind speed is needed to get to 30 psf; and vice versa, when the pressure coefficient is low, more wind speed is needed. That is, in monitoring and measuring for pressures reaching the critical level of concern, a large coefficient implies that the area is aerodynamically active and that the structure's geometry is stimulating the pressure, while a small coefficient implies that the area is aerodynamically protected and that a lot of wind is required to get the pressure to 30 psf.

Here probability gets factored into the mix. A climate model is generated, gathering all the available statistics of wind speed and direction for the site and formulated into a best approximation "probability mountain," with the peak of the mountain representing the higher chances of exceeding mild winds and the base of the mountain representing low chances of exceeding high winds. Probability is just an idealized way of talking about chance, so all of the data are examined to determine how many times—say, per year—the wind speed exceeds defined values from each direction. From any direction, the chance (or number of times per year) the wind will exceed 10 miles per hour is high (the peak of the mountain), and gets progressively less as the wind threshold gets faster. Also, certain directions are preferred—southwesterly winds are most common in many regions of North America—while others are not, especially for the stronger winds, and strong winds are naturally of most interest. A probability mountain, then, depicts contours of equal probability (see figure 31C)—the number of times per hour the wind speed, read from the circular gradations on the polar diagram, is exceeded. So 10-4 means that the wind speed is exceeded 1 hour in 10,000 hours.

Next, the wind speed contour for the response level of 30 psf is superimposed on the probability mountain like a cookie cutter (see figure 31D). Everything that falls outside the region of the cookie cutter contour is then added up, and that is the total probability of exceeding 30 psf from all directions. Luckily, this is a routine job for a computer, which can repeat the process for other response levels. Probability versus response is graphed to find the response corresponding to the probability of interest, say, once in fifty years. This process is then repeated for every point on the building for which measurements exist. Finally, contours are transferred to outlines of the building surfaces showing the expected fifty-year suctions (figure 31E). The designer can assess the situation at a glance and can adapt the building's cladding and glass accordingly.

Though simplified, this is the basic concept according to which all the wind tunnel data and all the wind information can be synthesized to provide the most realistic design information. The traditional method of taking the worst response and using it with the fifty-year return period wind speed always provides a more conservative response, and sometimes exceedingly so. In the above example, consider a location where the worst winds almost always come from directions other than those for which the high-pressure coefficients occur. By the traditional method, that location on the building would be defined for a situation that was expected to occur not once in fifty years but perhaps once in five thousand years.

Invented for the World Trade Center wind study, Davenport's probabilistic prediction method gained universal acceptance only twenty years later. Engineers are inherently conservative and cautious. Most are not trained in the mathematics of probability, and thus the lab devoted many conversations with clients to explaining the methodology. Similarly, most wind engineers do not arrive in the field via civil engineering but rather through fluid mechanics, or aeronautical or mechanical engineering. That was one of Davenport's great strengths. First and foremost he was a civil engineer. And he translated advanced methodologies formulated elsewhere into valuable tools for civil engineers. He was also a risk taker. He believed in analyzing risk rationally, assessing the probability of risk, and making decisions accordingly, even if the result flew in the face of tradition.

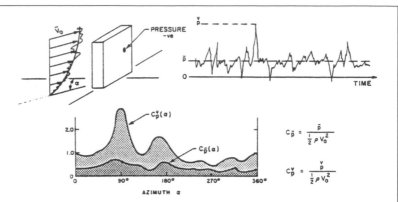

Figure 31A. Graphic displays showing what the lab gets from the wind tunnel. Clockwise from left: variation of wind speed with height; slice of a building with a pressure tap representative of a specific area under investigation, such as a window or a piece of cladding; time history indicating pressure variation; mean and peak pressure variation with wind angle. Courtesy of the Boundary Layer Wind Tunnel Laboratory.

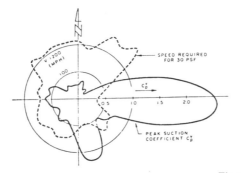

Figure 31B. Polar plots of the wind pressure coefficient (solid line) and wind speed required for 30 psf (dotted line). Courtesy of the Boundary Layer Wind Tunnel Laboratory.

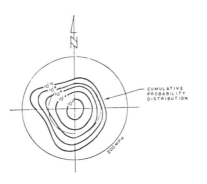

Figure 31C. A polar plot showing the annual cumulative probability distribution of gradient wind speed for New York City, defining the probability that the wind speed, in any angular sector, will exceed the various wind speeds indicated by the circular grid. The distribution is plotted as contours, which lend themselves to a three-dimensional mountain shape; hence this plot is also called a "probability mountain." Courtesy of the Boundary Layer Wind Tunnel Laboratory.

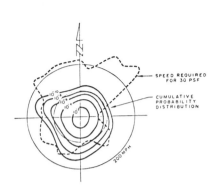

Figure 31D. The wind speed contour for the response level of 30 psf (the dotted line, the same as shown in figure 31B) is superimposed like a cookie cutter atop the probability mountain (the solid lines at center). The sum of all the probability outside the cookie cutter defines the total probability of exceeding 30 psf from all directions. Courtesy of the Boundary Layer Wind Tunnel Laboratory.

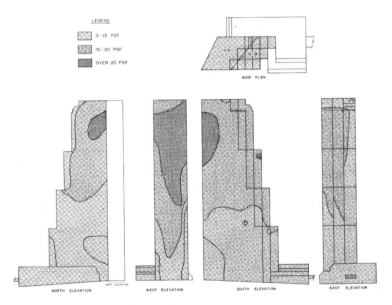

Figure 31E. Contours of exterior local peak pressures, predicted for 100-year return periods. Courtesy of the Boundary Layer Wind Tunnel Laboratory.

In this incessant tweaking of the towers' performance, they studied the effects of moving the towers closer together, farther apart, changing their orientation to one another, and adding damping. One tweak had Davenport rerunning the wind tunnel tests with the same models but rotating the service core of the southerly tower by 90 degrees in relationship to its outer perimeter. This made the structural properties of the two towers more asymmetrical, thwarting their inclination to oscillate like a tuning fork. It proved an easy and effective change. However, convincing the architect Yamasaki to make such a fundamental alteration at such a late stage in the game was another matter. "We told him in Detroit, and you could hear the screams in Los Angeles," recalled Robertson, though Yamasaki acquiesced in the end. Another successful tweak (also to the architect's chagrin) entailed removing one column along each side of the structures and widening the remaining columns. This produced an increase in stiffness, and in turn an increase in the structures' resonant frequency—raising the structures' frequency above that at which vortex shedding would occur, raising the towers above the flood, as it were, and thus decreasing opportunities for dynamic response.

Even with those fixes, the structures, though entirely safe as measured against their performance in the wind, still swayed more than might be prudent, considering the Oregon test results and people's tolerance for motion. This dogged Davenport until one day he walked into Robertson's office with an idea. "You know," he said, "if we could really get some damping into this structure, we wouldn't have to worry about the wind tunnel tests and things. The damping by itself would be very helpful." He knew about dampers in aircraft, used to muffle vibrations and smooth out the ride. He suggested filling the hollow columns of the structures with some kind of rubber to absorb the oscillations. This would be too complicated, thought Robertson. Instead he suggested friction dampers under the floor joists, and he proceeded to invent the device from whole cloth.

Not long after, he was in a meeting trying to convince the chairman of the board at 3M, the only company that could feasibly manufacture the viscoelastic meat that made the dampers work. "The application was outside of their ken," Robertson recalled. "[The

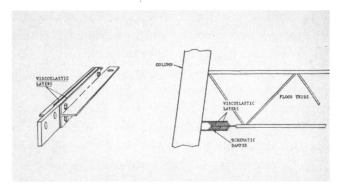

Figure 32A. An illustration of the viscoelastic dampers, designed by Robertson and custom-made by 3M. Courtesy of Leslie Robertson.

Figure 32B. Eleven thousand dampers per tower served as shock absorbers, reducing the towers' motion to within the realm of acceptability in terms of the human perception threshold. Courtesy of Leslie Robertson.

chairman] was concerned about liability. I explained to him that the buildings were perfectly safe without dampers. The dampers were just designed to reduce the sway component of the motion." The dampers, 11,000 of them per tower from the 7th to 107th floors, were essentially shock absorbers. Custom-made by 3M, they were double-decker structural steel sandwiches. The outer two layers of steel attached at strategic junctures to the towers' perimeter columns, the third inner layer of steel connected to joints in the flooring system, while in between the steel was the meat of the sandwiches, the movable viscoelastic material. These artificial joints, a medley of strength and elasticity, in a way reintroduced the innate structural creakiness, the natural give inherent in the dense masonry and riveted grid of steel girders that held up the Empire State Building. When the sandwich's outer steelwork swayed with the building, the interior layers of the dampers resisted: the relative motion between the bottom chords of the floor truss and the columns resulted in differential movements where the dampers were located. The resultant strain in the middle viscoelastic material dissipated the energy and dampened the sway. And while in the end, the maximum design deflection for the Twin Towers was in the neighborhood of 10 to 14 inches in 80-mile-an-hour wind, with oscillation period of 11 seconds, if they swayed even a fraction of that during the average gusty day, nobody was the wiser. The Port Authority of New York never received any complaints about motion, which led Robertson at times to wonder whether the ultimate design wasn't in the end too conservative, with all the diligent tests and the patent-worthy dampers. But the dampers did more than buffer the effects of wind. After the al-Qaeda attacks on the Twin Towers on 9/11, the dampers were credited with mitigating movement caused by the impact of the jetliners, preserving the integrity of stairwells and emergency exits and saving lives.

Davenport was stunned by the news of the attacks and the towers' collapse, which he heard on the car radio on the way to a medical appointment with Sheila. In his customary quiet way, he followed the news intently. At some point he spoke with Robertson, who was in Hong Kong that day. One month after the tragedy, attending a previously scheduled meeting of the nation's structural engineers, Robertson broke down and wept at the lectern when asked by a reporter whether he wished he'd done anything differently in the design of the buildings. Davenport had not participated in the planning stages' design work on redundancy, robustness, and aircraft impact, but after 9/11 he followed all ensuing reports and studies. At the lab, the model of the World Trade Center had always held pride of place, on display in the lobby. And there it remained, with an inscription: "The members of the Alan G. Davenport Wind Engineering Group wish to extend our deepest sympathies to the victims and their families of the tragedy, which occurred on Tuesday, September 11, 2001."

With the wind program for the World Trade Center, Davenport and Robertson had dealt with a broad spectrum of wind-related uncertainties empirically and effectively, thereby setting a precedent for all wind studies to come. In combining aerodynamic data on loads gleaned from the wind tunnel studies with wind climate modeling, Davenport could now provide reports with much more specific—that is, real—information. A project's structural engineer needn't worry about the wind climate or the building's aerodynamics because the wind engineer did all the worrying for him. Davenport's lab provided the data and the analysis, spelling out what the design responses

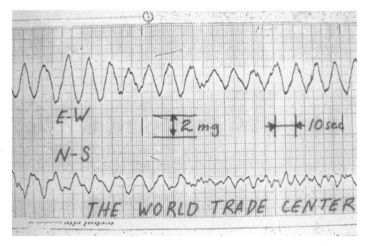

Figure 33. Recorded with accelerometers just below the roof of the World Trade Center's North Tower, these east-west and north-south traces measure acceleration (vertical on the graph; 2 milli-g is marked) versus time (horizontal on the graph; scale of 10 seconds is marked). The frequency of the tower in both the E-W and N-S directions is indicated at 0.1 Hz—that is, a period of about 10 seconds. The wind was exciting mainly the E-W motion at about 3–4 mill-g's, under whatever the wind speed was at the time. Courtesy of Leslie Robertson.

Figure 34. The towers of the World Trade Center stood still enough for tightrope walker Philippe Petit to traverse between their rooftops on August 7, 1974. As reported in the *New York Times*, "Combining the cunning of a second-story man with the nerve of an Evel Knievel, a French high-wire artist sneaked past guards at the World Trade Center, ran a cable between the tops of its twin towers and tightrope-walked across." He went back and forth eight times, more dancing than walking (as a police officer noted on his arrest). Petit hesitated only a moment in taking his first steps off the South Tower, recalling what someone had told him about the elasticity of structures and their ability to sway in the wind. He was undeterred, having conducted some wind tests of his own, practicing on a wire between two trees, exactly the distance between the two towers. As he recalled, "I asked my friends to make the cable dance." They shook and swung the wire between the trees with all their weight, trying to bounce him off in a rather boisterous if less empirical modeling of the wind. Courtesy of Leslie Robertson.

would be: the maximum pressures, forces, and deflections that the structure would endure with a once-in-fifty-years "design wind." Beyond the conceptual advantages, practically speaking this amounted to a 25 to 50 percent reduction in design wind loads.

Davenport's method, synthesizing the wind tunnel modeling data with the wind climate modeling data, also had the beneficial effect of interpreting the results clearly in the language of the design engineer. Previously, the crude smooth-flow wind tunnel tests provided engineers with reams and reams of data, and engineers would cherry-pick the worst-responding direction and use that with the design wind speed, no matter what the probability of the occurrence of such a wind. With Davenport's method, the lab factored in the same mass of detail and data but provided clear conclusions in language the design engineer could understand—providing, for example, a graph of the response versus return period, or contour maps of fifty-year return period pressures on the surface of the building, with simple multiplying factors to calculate results for different return periods if need be. Substantively and logistically, these were pioneering achievements, milestones that moved the field beyond the dark ages, beyond the guesswork of the past, providing the inspiration—and the template—for future investigations.

The actual template was nothing more complicated than a clean conceptual equation, distilling Davenport's knowledge about the ways of the wind into a format that anyone could understand: *wind load = wind climate + influence of terrain (roughness, topography) + aerodynamic response + mechanical response (stresses, deformations, accelerations) + criteria (strength, deflections, comfort)*. Analogically, he called it the "wind loading chain" (see sidebar, "Davenport's Wind Loading Chain," below). With the World Trade Center

Davenport's Wind Loading Chain

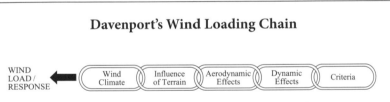

Figure 35. Davenport's analogical tool. Courtesy of the Boundary Layer Wind Tunnel Laboratory.

In July 2011, the General Assembly of the International Association of Wind Engineering (IAWE) unanimously approved use of the term "Alan G. Davenport Wind Loading Chain" for describing Davenport's approach to evaluating wind loads and wind-induced responses for buildings and structures. The lab's consulting director, Nick Isyumov, had put forward a proposal advocating the roots and relevance of Davenport's handy conceptual analogy. The statement read in part:

"Alan Davenport's doctoral research laid the foundation of today's wind engineering. His approach was based on the chain of thought which recognized that the wind loading experienced by a particular building or structure is determined by the combined effects of the local wind climate, which must be described in statistical terms; the local wind exposure, which is influenced by terrain roughness and topography; the aerodynamics characteristics of the building shape; and the potential for load increases due to possible wind-induced resonant vibrations. He also recognized that clear criteria must be in place for judging the importance of the consequences of the predicted wind action. This included the effects of wind on the integrity of the structure and the exterior envelope and various serviceability considerations, such as the control of the wind-induced drift, the effects of building motions on occupants and the usability of outdoor areas at and near particular buildings and structures. In his papers he referred to this process of evaluating the effects of wind action as the 'wind loading chain'. This was in recognition that the evaluation of the wind loading and its effects relies on several interconnected considerations, each of which requires scrutiny and systematic assessment. With analogy to a physical chain, the weakest link or component in this process determines the final outcome. Little is gained by embellishing strong links but much is lost by not paying attention to the weak ones.

"Not only does the wind loading chain provide a firm basis for the evaluation of wind loads, it also forms the basis for the chain of thought needed to define and solve most other wind engineering problems. The format of Alan's wind loading chain has been adopted in the specification of wind loads in most building codes and standards worldwide."

study he not only set this template for the overall chainlike process, he also solved the worst of the conceptual problems encountered at every link. As he put it into play over the course of his career, the chain continued to evolve. But he always preserved the simplicity, Davenport's signature simplicity being one of the main reasons why his approach to solving wind problems caught on.

ψ   ψ   ψ

While working on the World Trade Center, Davenport taught all his classes on Mondays, traveled to Colorado or Seattle or New York for the rest of the week, and returned home on Fridays (his daughter Anna once mentioned to a friend, "My dad goes to work in an airplane. How come your dad goes to work in a car?"). Jet-setting had its hassles, and the pace took its toll, even on the stoically imperturbable Davenport. Once, before boarding the plane, he used part of his travel allowance to purchase half a million dollars worth of insurance at an airport kiosk. The insurance was not for himself but for the delicate model of the World Trade Center that he carried aboard. Purchasing insurance onsite at the airport was not unusual practice at the time. But nonetheless, this behavior attracted undue attention. The pilot expressed concern about this package, roughly the size of a bomb, which Davenport had gingerly stowed beneath his seat. There was a commotion. Departure was delayed until the item was opened, inspected, and explained. Even then, Davenport sensed he was under surveillance for the rest of the journey.

Seeking to minimize such hassles, and, more to the point, desiring full creative freedom, Davenport dreamed of building a wind tunnel all his own. One Sunday, he wangled a seat at a luncheon meeting held by the University of Western Ontario's board of governors at board chair D. B. Weldon's farmhouse. Davenport had a core of steel when he set his mind to something, and with that mindset he made his pitch, and the deed was done. The board decided that the university would provide accommodation for a wind tunnel, though not in the neogothic, ivy-clad limestone buildings that were the hallmark of the campus. Rather, the wind tunnel would reside temporarily in a more utilitarian prefabricated metal structure next to the engineering building. With a $30,000 National Research Council grant, construction was completed in six months (and this "temporary" structure lasted forty years). In November 1965, Davenport's Boundary Layer Wind Tunnel Laboratory opened officially in a ceremony made complete by festive pinwheels and a front-page photograph in the *Globe and Mail*.

Figure 36. The official opening of the boundary layer wind tunnel laboratory, November 1965. Left to right, University of Western Ontario president George E. Hall, National Research Council vice-president K. F. Tupper, Dean of Engineering Richard Dillon, and Davenport. Courtesy of the Boundary Layer Wind Tunnel Laboratory.

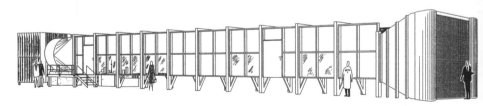

Figure 37. Producing winds reaching 34 miles per hour, the main components of the boundary layer wind tunnel included (from left) the 35-horsepower fan, a 90-foot test section, and the inlet and contraction. Courtesy of the Boundary Layer Wind Tunnel Laboratory.

Figure 38. Nicolas Isyumov, assistant director and manager during the lab's early days, in the fall of 1965, looking into wind tunnel's test section from the control room. Since the lab still did not yet have its state-of-the-art instrumentation in place, the "manometer board" on view is a prop. In an era before electronics and computers, the mean pressures were measured by a manometer board, with tubes traveling from the wind tunnel model into the back of the board. The air pressure in the tubes displaced the manometer's reservoir of red-dyed water, thus registering the levels of pressure along the board's scale with red bars (at center), and the bars were then measured and recorded with pencil and paper. Courtesy of the Boundary Layer Wind Tunnel Laboratory.

No sooner was it generating turbulent wind flow than Davenport's laboratory attracted as much research as it could handle. The work came fast and furiously, and for a few years Davenport turned to cigarettes to keep pace, filling his office with clouds of smoke. One of the first projects the new lab took on was a subsequent World Trade Center study, analyzing airflow at pedestrian levels in the plaza. In the very early days Davenport and his associates provided expert testimony in the lawsuit following the collapse in high winds of three cooling towers at the Ferrybridge Power Station in England, a disaster that was nothing short of Europe's Tacoma Narrows. The lab also conducted a broad study of potential geometric forms for the U.S. Steel Tower in Pittsburgh. Of the six cross sections tested—circle,

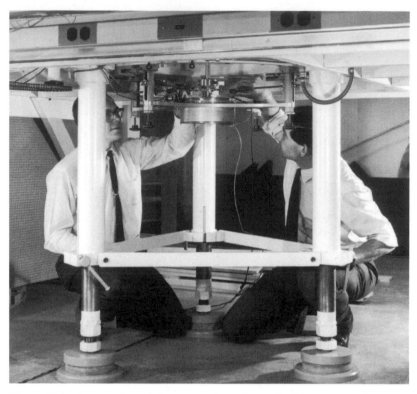

Figure 39. Isyumov and research associate Peter Rosati (right) inspecting the underbelly of the test section's turntable on which the targeted model sits. Courtesy of the Boundary Layer Wind Tunnel Laboratory.

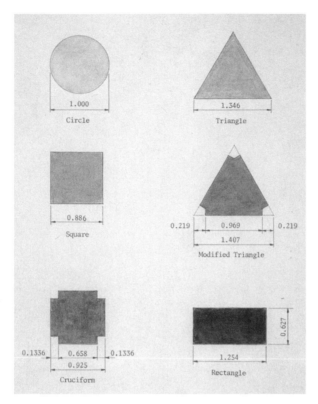

Figure 41. The lab conducted a survey study of potential geometric forms for the U.S. Steel Tower in Pittsburgh. Courtesy of the Boundary Layer Wind Tunnel Laboratory.

Figure 40 (Facing page bottom). Following the November 1965 gale-induced collapse of three cooling towers at the Ferrybridge Power Station in Yorkshire, England, the lab ran precautionary tests for the American Electric Power Service Corporation. Pictured here is a model of the Fort Martin Cooling Tower in West Virginia. Davenport and associates had provided expert testimony in the lawsuit following the Ferrybridge collapse, a disaster that was nothing short of Europe's Tacoma Narrows. For the American Electric study, the lab pioneered the use of Devcon, an epoxy material that had the same density as reinforced concrete, and from which aeroelastic models could be constructed for structures such as cooling towers. Performed in 1966, these tests produced the first formal report issued by the newly minted lab. Courtesy of the Boundary Layer Wind Tunnel Laboratory.

Figure 42. The U.S. Steel Tower ultimately took the shape of an aerodynamically superior triangle. Courtesy of the Boundary Layer Wind Tunnel Laboratory.

square, cross, rectangle, triangle, and a modified triangle (with the corners lopped off and replaced by a V-shaped notch)—the circle came out on top. This was not surprising. Structural engineers knew full well what basic shapes were more or less aerodynamic. Alas, dictating the optimum shape was not the engineer's place. The engineer's role was more to facilitate the plans, the whims, of architects and owners. Backed with the scientific credibility generated in a boundary layer wind tunnel, however, engineers could leverage more influence on design. In the case of the U.S. Steel Tower, it ultimately took the shape of a triangle—superficially appropriate, insofar as it

Figure 43. The pressure model for the U.S. Steel Tower was equipped with a "scanivalve" inside the model to ensure short tubes and hence high-frequency response. The scanivalve had forty-eight ports, each of which could connect a pressure tap on the model's surface to a single embedded pressure transducer, and the transducer's electrical signals were then measured by the PDP-8 computer for the full-scale equivalent of about an hour (20 to 60 seconds), before stepping to the next port Courtesy of the Boundary Layer Wind Tunnel Laboratory.

was to be located in the downtown Pittsburgh district known as the Golden Triangle. Wisely, however, the owners opted for the slightly more aerodynamic of the two three-sided polygons proposed, the triangle with indented corners, whose notches served as geometric speed bumps, increasing the levels of aerodynamic damping and ameliorating the influences of the wind. And as ground was broken in August 1966 for the World Trade Center, on target to claim the title of world's tallest towers, Davenport found himself involved with

Figure 44. The helipad atop the U.S. Steel Tower also underwent a rigorous regime of wind tests, with smoke used to visualize the zone of influence affecting the operation of the aircraft. Courtesy of the Boundary Layer Wind Tunnel Laboratory.

two new projects that were shooting to be taller still. Indeed, earlier the same year, the *Engineering News-Record* had reported an outbreak of "skyscraper fever."

Business was booming for Sears, Roebuck & Company in Chicago, and designs were in the works for new headquarters that would practically and symbolically accommodate the company's success. Sears wanted to move its seven thousand employees from the suburbs into the Windy City's downtown core. The structural engineer was Fazlur Khan, "the Einstein of engineering," a gifted visionary genius about whom colleagues could never string together too many superlatives. In eulogizing Khan at a posthumous tribute (he died suddenly of heart failure in 1982, at age fifty-two), Davenport borrowed a quotation from the Koran: "I will set my face to the wind and scatter my seed on high."

For the Sears Tower, Khan and architect Bruce Graham, both of the Chicago firm Skidmore, Owings & Merrill, toyed with a number of options to satisfy their client's palatial demands—two million square feet for Sears's own use and two million for rental office space, with a budget of $200 million. They considered a tiered "wedding cake" design, with the bulk of the volume at the base. They pondered two separate buildings. The favorite was an ultra-tall, slender structure, the only disadvantage being that the slender silhouette demanded a modest decrease in overall floor space. In persuading the owner that this was the premium design, and that such a tall structure would not cause a proportionate spike in price, Khan presented the client with his trademark charts showing the correlation between building height and various structural systems. His message was simple: there need not be a significant economic penalty for having a tall building as long as the proper structural scheme was employed. Sears took Khan's counsel, accepting an architectural program for a "bundled tube" or "modular tube" creation whereby nine mini-towers of differing heights rose with staggered rooftops, the overall structure tapering up to 1,451 feet. By 1969 the project was ready to proceed. At this stage in his career, Khan, the structural chief and a partner at SOM, rarely involved himself in the engineering minutiae unless problems arose. It soon became apparent that the wind analysis for the Sears Tower would benefit from his continued involvement. Two issues needed tending to. What wind pressures would his eccentrically asymmetrical tower attract? And how would the tower respond?

Not five years earlier, Khan had attempted to address the same questions with Chicago's 1,125-foot John Hancock Center. With no boundary layer wind tunnel yet in existence, he resorted to an aeronautical wind tunnel, where the smooth flow analysis was rudimentary, but better than nothing. He also went to great lengths to devise a cheap and practical way of testing the scale of motion that could cause discomfort to the tower's occupants. During a Sunday family outing to Chicago's Museum of Science and Industry, he found inspiration in a washing machine exhibit featuring a large, clear-walled tub positioned on a rotating podium. Standing on the turning platform, Khan discerned a jerk, which sparked an idea: this turntable could simulate the jerk of swaying buildings. He inquired of the museum and successfully co-opted *The Tale of the Tub* exhibit for his own purposes, which is to say he conducted a rather makeshift experiment and developed a general "threshold of discomfort" similar to Davenport and Robertson's. Khan corroborated his results by measuring the discomfort experienced in a Chicago apartment building known for its noticeable sway (using as his guinea pig a former student, Phyllis Lambert, the Canadian philanthropist who in 1979 founded the Canadian Centre for Architecture in Montreal). But these were crude conclusions at best. To err on the side of caution,

Figure 45. The Sears Tower in Chicago boasted an innovative "bundled tube" design, with nine mini-towers of staggered rooftops tapering to reach 1,451 feet. Nina B./Shutterstock.com.

Khan ultimately decided to implement design wind pressures 25 percent greater than those recommended in the Chicago building code.

By the time the Sears Tower was on the boards, Khan had gamely retained Davenport's lab to conduct more credible tests. Davenport's presence around the SOM offices made a marked difference in the professional atmosphere. In the face of stressful situations—making substantive changes to multi-million-dollar architectural plans—Davenport maintained his preternatural calm. And he became known for his sophisticated analytical mind. He was not merely a technician, though he was a great technician. He also offered valued input on design issues—himself keenly aware that such input would be a very important step forward for wind engineering to take, to be more than a technical adjunct to the design, to be a crucial contributor. In this respect, the Sears Tower, like the Twin Towers, was precedent setting in that it introduced wind as an integral factor early in the design process. Wind engineering analysis was no longer merely an afterthought.

Figure 46. At the Symposium on Tall Buildings at the University of Southampton, 1966. Seated left to right, the Sears Tower's structural engineer Fazlur Khan of Sidmore, Owings & Merrill, Liselotte Khan, Betty Stafford Smith, Davenport, and Les Robertson; standing left to right, Todor Karamanski and Bryan Stafford Smith. Courtesy of Sheila Davenport.

The most significant difference between the World Trade Center and the Sears Tower projects was the latter's irregular shape. It was not a constant prismatic form. Granted, the base was a perfect square, with a footprint of three-by-three square tubes. Halfway up, however, two corner tubes fell back, leaving seven. A ways up again two tubes begged off, leaving five in a cruciform shape. And from there two tubes broke away for the peak. This asymmetry made the building's behavior much more complicated. The testing for the rectilinear Twin Towers had required only a two-degrees-of-freedom or "stick" aeroelastic model—simulating the wind-induced dynamic response in only two fundamental sway modes, rotating about the classic x-y-axes, for both the along-wind and across-wind directions. The Sears Tower required a model boasting multiple degrees of freedom, with an additional axis of twist accounting for its torsional vibration. And since each of its three axes had several vibrational modes excited by the wind, the model had to provide 21 degrees of freedom, seven in each of the vertical, horizontal, and torsional axes.

A model with 21 degrees of freedom represented quite a wind engineering feat. The gadget had seven lumped-mass rigid-floor diaphragms; it had flexible rod-column elements to mimic the axial deformation; and it sat on a flexible base, reflecting the slight rotational give of the foundation itself. Discontinuous patches of balsa wood clad its exterior—discontinuous so as to prevent the cladding from inadvertently enhancing the structure's stiffness. The building's capacity for self-damping, based on conservative estimates due to a dearth of data, was ingeniously mimicked by applying a viscous coating to the model's columns. Rendered with great care down to a hundredth of an inch, this fine engineering specimen took a year to design and construct. The model, in the end, was a technical wonder, a work of art. All the same, this cutting-edge scientific instrument might have belied its own sophistication to the average observer. As noted by John Zils, the project manager and now an associate partner at SOM: "It looked like a Rube Goldberg machine"—referring to the Pulitzer Prize–winning cartoonist known for his creations depicting overengineered devices performing basic tasks in absurdly convoluted ways.

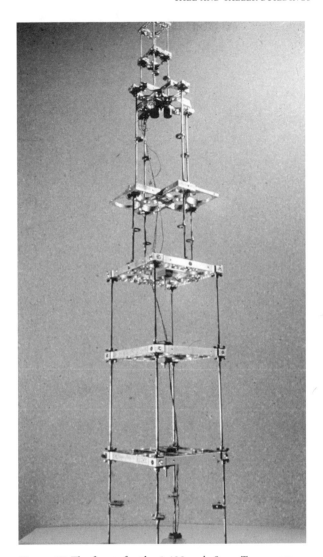

Figure 47. The frame for the 1:400 scale Sears Tower aero-elastic model, which allowed for an unprecedented 21 degrees of freedom spread over seven levels. Courtesy of the Boundary Layer Wind Tunnel Laboratory.

Figure 48. Upstream view of the Sears model in the wind tunnel. Discontinuous patches of balsa wood clad the model's exterior—discontinuous to prevent the cladding from inadvertently enhancing the structure's stiffness. Courtesy of the Boundary Layer Wind Tunnel Laboratory.

ꙮ   ꙮ   ꙮ

When the Sears Tower model was first exposed to the wind tunnel's analogue to reality, Khan confronted some unexpected results. He had predicted that the highest wind pressures would be near the top of the building's 108 floors, while tests showed the strongest pressures were in the vicinity of the 50th and 66th floors, prompting extra cladding supports planned for the peak to be moved to the midsection. More alarmingly, the overall prognosis for the tower's structural integrity did not look good. The tower crossed what is called the "threshold of aerodynamic instability"—given winds from certain directions at certain speeds, the tower would become unstable and unsafe. To buttress his tower, Khan considered stiffening the service core, or the tubular core, but either would compromise the interior layout to an unacceptable extent. He also considered mechanical damping, and flew to New York to inspect the dampers being installed in the World Trade Center. That option, he decided, was too pricey and too risky, since it was new technology and thus a bit of a wild card in terms of its long-term viability.

While weighing the alternatives, Khan directed Davenport to continue testing, specifically, to refine his statistics modeling the Windy City's wind climate. He wanted the wind analysis to be as cutting edge as the model. And from these efforts came the solution. Within weeks, Davenport's evaluation proved that the recurrence period of winds causing the tower to cross the threshold of instability was incredibly high—they would occur very infrequently, once in 1,000 years or more. Here directional analysis played a crucial factor, with the worst instability occurring only at very rare wind directions. The design speed criteria specified that the structure must be able to withstand a much lower, 100-year-recurrence wind, which itself was above and beyond most building codes and standards (design speed is not be confused with failure speed; exposure to winds at design speed should not cause any damage, while at failure speed there would probably be catastrophic failure; safety factors meant that the building would in fact perform safely for a 500-year wind). During the decade following the Sears Tower wind tests, the lab extended its intensive analysis of Chicago's wind patterns even further and formulated a statistical model of the wind climate, correlating the study results with more and more full-scale data. This in turn served as justification to update the Chicago building code.

It might seem counterintuitive that Davenport's more scientific and sophisticated approach to wind engineering would have the effect that architects and engineers could at times be less conservative in their preventive measures against wind forces. Davenport's modus operandi, in the lab and in life, was based on the principle of informed risk. This was often much to his wife Sheila's dismay—he encouraged their sixteen-year-old son, Tom, to cycle what might seem a risky 500 miles to Montreal to take in the Olympics in the summer of 1976; and he didn't think twice about their teenage daughter, Clare, backpacking around Europe, even when most of her friends' parents rejected the idea outright. His intuition for calculated risk was honed during his pre-university actuarial training and turned

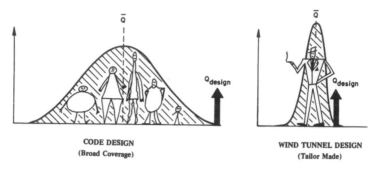

CODE DESIGN
(Broad Coverage)

WIND TUNNEL DESIGN
(Tailor Made)

Figure 49. Davenport liked to emphasize that designing a structure based on code recommendations was a one-size-fits-all approach, whereas designing based on wind tunnel results produced a much better, bespoke fit. Courtesy of the Boundary Layer Wind Tunnel Laboratory.

out to be useful in assessing problems of risk in engineering. The study of risk was always top of mind—perceived risk versus real risk, the assessment and management of risk. More data meant more focused decisions, based on a more accurate representation of reality. And it avoided the need for precautionary concessions driven by better-safe-than-sorry guesstimates.

But then again, more nuanced investigations just as often exposed unexpected risks that were too risky to take. This was the case with the wind program for Toronto's CN Tower. It was in the queue at the Davenport lab concurrently with the Sears Tower, and it too produced a very progressive model—for very different reasons, and with very different results.

At 1,815 feet, the CN Tower would cinch the title of the world's tallest freestanding structure (as opposed to an occupied building, for which more stringent wind design standards applied). When completed, it would exceed the height of the Moscow Ostankino Television Tower by some 164 feet. It would climb well beyond the highest building in Toronto at the time, a mere 56 stories and 732 feet. Even at that meager maximum, the city's hitherto squat skyline had developed a respectable silhouette. Some of its skyscrapers caused trouble for the existing transmission towers, with signals bouncing around, ricocheting back and forth between buildings, blurring television and radio reception. The world's tallest freestanding structure, with microwave receptors at 1,109 feet and at the very tip of the antenna, would solve these problems by providing a futuristic communications hub.

The master site plan, drawn up in 1967, was initially known as Metro Centre, a large land redevelopment project repurposing a railway switching yard between downtown and the waterfront. Financing came from both the Canadian National (CN) and Canadian Pacific (CP) rail companies. The tower provided the pièce de résistance of the complex, with a corresponding price tag. Owing to escalating budgets, CP backed out, leaving CN to go it alone. Digging began in 1971, but by then the complex had been dismembered into disparately administered elements, and it progressed piece by piece—a bridge here, another bridge there, and the tower, designed by the expat Australian architect John Andrews together with the Toronto-based firm WZMH Architects.

Urbanistically, the entire Metro Centre plan was a disaster, according to Franz Knoll of the Montreal-based firm Nicolet Chartrand Knoll Engineering Ltd., the lead structural engineer on the project, who also has the Louvre Pyramid on his résumé. To make matters worse, consensus had it (among the engineers, at least) that the original design for the world's tallest freestanding structure was as ugly as sin. It was considerably stouter than the tower ultimately erected, sitting on not one but three separate legs linked at various heights by bridges. And not only was it an eyesore, its cost kept increasing. An architect could get away with nearly anything in the name of art. But the escalating expense was the coup de grâce for this tripodic behemoth. The redesign produced an economically, structurally, and aesthetically superior landmark, elegant in its simplicity. From a Y-shaped base, three graceful supports were melded into a single continuous hexagonal core, rising to the bulbous Space Deck at 1,467 feet. The simplicity of the structure allowed Davenport to devise a model that, from a technical point of view, was more than just an analogue or equivalent form. It was an exact replica—the model of the tower's concrete shaft had exactly the same continuous structural geometry as its full-scale twin. This exactitude was not just a happy by-product of the structure's simplicity, it also served a practical pur-

Figure 50. Standing with a partial model of the CN Tower, the lab's various directors over the years. From left, Barry Vickery, David Surry, Nicholas Isyumov, Milos Novak, Davenport, and Peter King. Courtesy of the Boundary Layer Wind Tunnel Laboratory.

Figure 51. The lab also ran snow tests on one of the six competitors for the domed stadium that would sit nestled near the CN Tower's base. Courtesy of the Boundary Layer Wind Tunnel Laboratory.

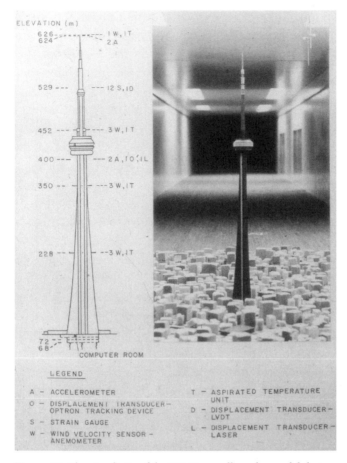

ELEVATION (m)

626 ──── I W, IT
624 ──── 2 A

529 ─── ─── 12 S, 1D

452 ─── ─── 3 W, I T

400 ─── ─── 2 A, IO'IL

350 ─ ─ ─── 3 W, IT

228 ── ─ ─ ─3 W, IT

72
68 ──── COMPUTER ROOM

LEGEND

A — ACCELEROMETER

O — DISPLACEMENT TRANSDUCER—
OPTRON TRACKING DEVICE

S — STRAIN GAUGE

W — WIND VELOCITY SENSOR —
ANEMOMETER

T — ASPIRATED TEMPERATURE
UNIT

D — DISPLACEMENT TRANSDUCER—
LVDT

L — DISPLACEMENT TRANSDUCER—
LASER

Figure 52. The simplicity of the CN Tower allowed a model that
was more than just an analogue or equivalent form, it was an exact
replica. Courtesy of the Boundary Layer Wind Tunnel Laboratory.

pose. The CN Tower had to be tested more extensively than either
the Twin Towers or the Sears Tower because, while its overall design
was simple, it had intricate, finicky components that demanded rigorous evaluation.

The replica model made it possible for the engineers to selectively
attend to the most important motions induced by wind. "Wind engineers are pragmatic and usually must do with limited resources," explains Nick Isyumov, the project director, now a consulting director

at the lab (Isyumov made seminal contributions in the area of snow loading, among others). "Consequently, standard aeroelastic models are designed to provide information on those parts of the wind-induced response which are judged to be important—important things are modeled, unimportant things are not."

Both the standard analogue model and the replica model exactly represented the exterior geometry in order to capture the external wind forces. But the replica also reproduced all the tower shaft's innards. "The model correctly simulated *all* wind-induced responses of the prototype tower shaft," says Isyumov. "This included flexural actions of the shaft, as well as its wind-induced rotation or twisting about the vertical axis. Remember, the tower shaft had three tapering 'legs' positioned at an angular spacing of 120 degrees. The shaft is symmetric for wind directions in the plane of these legs, but is unsymmetric for other directions. Hence the shaft was expected to twist as well as bend under the action of wind." Producing the replica was a challenge, as even the material properties had to be substantively duplicated to scale. For instance, sheets of Devcon A, a metalized epoxy used by plumbers, stood in for the concrete, precast to the correct thickness to serve as the tower's core.

The tower's list of wind issues went on and on. The transmission equipment was a devilish culprit. Subjected to the elements—freezing rain, for instance—the equipment would be irreparably damaged or at the very least temporarily dysfunctional. This posed a problem of how the equipment could be protectively enclosed. A steel box would interfere with broadcast signals. An air-supported bubble, a "Radome" built by Birdair, a leader in the field of tensile fabric architecture, seemed to fit the bill. The Radome, however, was a floppy, lightly pressurized bubble that posed its own problems. If it deflated or even dimpled in the wind, the membrane would come in contact with the transmission dishes, again causing interference with signals, damaging equipment, or tearing the bubble's sheeting. It fell to the lab to figure out the optimal pressure needed to keep the bubble inflated. The more pressure, the more resistant the bubble would be to dimpling, but too much pressure and the material would degrade prematurely.

ധ   ധ   ധ

The lab troubleshooted similar issues with two related forays into "tensile membrane technology."

The long-span tensile membrane roof of the Calgary Saddledome, which became as much an architectural icon for this western metropolis as the CN Tower did for Toronto, took the shape of a hyperbolic paraboloid, though its perimeter plan was nearly circular. The "membrane" was made from 24-inch-thick, precast concrete plates, suspended by post-tensioned cables—a mesh of steel cables, with double-sided concrete frosting. The roof's thickness was roughly 1/250th of its length, meaning the concrete would be flimsy, not providing all that much stability. Its response to the wind would be like that of a hammock erratically swaying back and forth, up and down, as two occupants plopped in and tried to get comfortable. The lab's study focused on the nature and magnitude of these fluctuating loads, in particular the out-of-balance, asymmetrical loading patterns, the havoc they could wreak, and their unwieldy potential for dynamic amplification.

Figure 53. The roof of Calgary's Saddledome—a "membrane" of 24-inch-thick precast concrete plates, suspended by post-tensioned cables—responded to windy gusts much as a trampoline responds to energetically bouncing children. Courtesy of the Boundary Layer Wind Tunnel Laboratory.

Davenport developed novel experimental and analytical techniques to measure the dynamic loads driving all the important modes of the Saddledome roof vibration. These techniques made use of what's called an "influence surface," a mathematical analytic tool that describes distributed loads on broad surfaces by collapsing and combining a myriad of forces, integrating them into a few generalized shapes. "Influence surfaces—or one-dimensional influence lines for linear structures such as towers or bridges—simply relate the effect at one location to loads applied elsewhere," explains David Surry, one of the lab's consulting directors who over the course of his career there made fundamental contributions in the field of low buildings. In addition to a hammock, Surry imagines the Saddledome roof as a large trampoline, with kids bouncing around like the wind dancing over the roof's surface. "The simple up-and-down motion of the trampoline is sensitive to the loads acting in the same direction all over the trampoline," he says, "with higher weighting for those near the center, whereas for a sway mode like the hammock example, the important aspects of the dynamic loads are those that act up on one side and down on the other." By statistically summing the effects from all the important sway modes of vibration, Davenport obtained the total dynamic response of the roof when excited by the wind, which he in turn translated into an equivalent static load (more like the stationary load caused by snow). "The aim in many of our test procedures is to provide the designer with simplified information," says Surry. "Most engineers design structures for wind using static loads. That's how codes work. In fact, most designers are not comfortable with dynamic concepts and unsteady loads of any kind, let alone stochastic ones. An effective static load is simply that static load which will reproduce the worst dynamic load effects."

The lab went on another foray into tensile membrane technology with the fabric roof designed for the Hajj Terminal in Saudi Arabia. Designed by Fazlur Khan for the King Abdulaziz International Airport in Jeddah, the structure emulated a Bedouin desert tent and was intended to serve as a shelter for the influx of pilgrims en route to Mecca—a projected one million people annually by the target design year of 1985. The open-air structure, with no walls, was to

Figure 54. Davenport with technician Rick Allen (right), admiring the Hajj Terminal model, now in the Pompidou Centre in Paris. Courtesy of the Boundary Layer
Wind Tunnel Laboratory.

Figure 55. In the wind tunnel the Hajj Terminal model displayed no instabilities,
but quivered and shimmered like Jell-O. Courtesy of the Boundary Layer Wind
Tunnel Laboratory.

Figure 56. The Hajj Terminal can accommodate 80,000 travelers at a time. Courtesy of SOM | Jay Langlois © Owens Corning.

cover an area of more than four million square feet, sectioned into 210 tent units, each unit 150 by 150 feet with a fabric roof rising to an open truncated peak. Embarking on the wind study, Khan inquired of Davenport, "What do you think the process will be on this?" To which Davenport replied, with keen enthusiasm, "I haven't a clue!"

The wind tunnel studies for the Hajj Terminal were the first tests on fabric roofs, generating a masterpiece of a model, which currently resides in the permanent collection of the Pompidou Centre in Paris. Unlike the CN Tower's air-supported fabric bubble, the Hajj Terminal uses a cable-supported tensile system, the tension emanating from radial cables sewn into sleeves on the interior and stretched into a parasol-like shape. Researching materials for the architectural model, SOM engineers found themselves shopping in the sewing section of Marshall Field & Company, trying to find a suitably stretchy fabric.

Once at the lab, the modeling had to be a tad more precise. The job went to one of the young new creative talents, research engineer Bjarni Tryggvason (later one of Canada's first astronauts). Tryggvason had been hired just the year before, following a poetic letter of application that appealed to Davenport's philosophical inclinations. It opened with "The beauty of the world is a never ending amazement," and continued:

> The opening of a rose petal is an often quoted example. And there are many more; the stars at night, the red glow of a sunset, a rainbow shining through a warm drizzle. But these are superficial beauties. Beauties that all can see and admire. It is when one delves into the causes of these that a much grander beauty emerges, one veiled in a cloak of intricacy and, as yet, mystery. We begin to see how all the minor events fit into the large scale occurrences of our everyday lives, and how each event, no matter how insignificant it is in appearance, plays a vital role in the way of the world.

Two paragraphs later, he asked,

> But what is it to understand? Is it simply the acquisition of the ability to describe events through mathematical relations? No. A true understanding must go beyond that. To my present mind, we can, through technical methods, understand only to within some finite limit. Beyond that, our understanding must be attained at an intuitive level of an open mind. I think this is possible and it is my goal.

Davenport, recognizing a kindred spirit, hired him without a second thought. And for the Hajj project, Tryggvason was charged with the somewhat poetic but mostly tricky task of modeling the tents' roof fabric. On the real-life structure, the fabric was fashioned from Teflon-coated fiberglass manufactured by Owens-Corning. Davenport visited the Owens-Corning plant and tested a mock-up by bouncing on it as if on a trampoline. After some shopping around, the model-sized replica was fashioned from polyethylene 1/50th of an inch thick, made to order by DuPont's research labs. Next came the chancy task of deforming the fabric into the appropriate shape with heat, chancy because the optimal molding temperature of the polyethylene was perilously close to its melting point. If the molding temperature were too low, however, the polyethylene would refuse to hold shape. In his quest for temperature control, to within three degrees of about 180 degrees Celsius (roughly the temperature for baking a cake), Tryggvason equipped the funnel-shaped molds with their own custom coil heaters, and he built a bespoke oven—a six-by-six-by-two-foot furnace with a wood frame, plywood siding, and lined with 10 inches of insulation. This all went beyond the purview of regular engineering practice. On the occasion of the lab's annual summer party that year, Tryggvason baked a three-layer cake in the Hajj's tent mold, with the cables rendered in decorative icing.

After a few trial runs, Tryggvason and a student, much to their own surprise, finished the model over a single weekend. Into the wind tunnel it went. Anxious expectation had it that some unusual behavior—perchance even flapping-flag-like instabilities—was likely to occur. But for all the worry about the tents' elastic response, within half an hour the testing proved anticlimactic. The only exceptional effect noticed during the wind tests was that the Hajj's tents shimmered like a bowl of Jell-O. There were no instabilities. The tests confirmed that Khan's architecture would provide a stable abode, even under strong winds. And in fact, the structure's geometry generated an optimal airflow pattern, organically expelling hot air through the central vents rather than allowing it to sit still and stagnant, thus providing a pleasant environment for the transient village that would reside beneath.

With the CN Tower, larger-scale tests on the air-supported Radome, at a ratio of 1:60, determined that if the Teflon-coated fiberglass-rayon fabric was inflated to five times normal and maintained at a constant pressure, it would effectively withstand the design winds (and with this exaggerated inflation, the balloon-like structure's skin, measuring only 1/32nd of an inch thick, could also hold the weight of an average adult man). Yet the wind issues for the CN Tower were still not all laid to rest. The partially open-air elevator shafts posed another hurdle, as did the panoramic expanses of windows, including the thick glass floor of the Sky Deck.

And at the very peak there was the antenna. A structure on top of a structure, it had 44 pieces standing 335 feet, nearly one-fifth the tower's total height. Thus, in addition to testing the antenna as part of the full-structure tests, the lab modeled the antenna on its own. Since its behavior in the wind was more predictable, with no torsional twisting, this model didn't require nearly as much fiddly precision as the shaft's model. A more makeshift creativity sufficed, involving the repurposing of hypodermic needle tubing. And while test results did not show any issues of concern, Davenport and his team recommended that the antenna be equipped with two tuned mass dampers(for another example, see the sidebar, "One Tower, Two Tuned Mass Dampers," page 98). The first tuned mass damper had been invented in 1909 by one Herman Frahm, who received U.S. Patent No. 989,958, "Device for Damping Vibrations of Bodies," which described the invention as a vibration cancellation device in which an auxiliary body is arranged within or on the structure in question to counterbalance and cancel oscillating forces (the purpose was to reduce the roll of ships and ship hull vibrations). The CN Tower antenna's damper system as devised by Davenport was passive, wobbling in response to the structure's movement and cancelling out the energy. It consisted of two doughnut-shaped steel rings (approximately 10 feet in diameter, 15 inches wide, and 12 inches deep), each ring containing nine metric tons of lead and attached to the antenna mast with supportive steel beams and universal joints, allowing the rings to pivot slightly. As the antenna deflected when exposed to wind excitation, the weighted rings offset the motion, a bit like a Hula-hoop swiveling around a gyrating waist.

One Tower, Two Tuned Mass Dampers

The 1,014-foot Sydney Tower was the first building designed to be equipped with a tuned mass damper. At the design stage in the 1960s, the structural engineer realized that the circular stack-like tower would have problems with vortex shedding in the first mode vibration. The main column had been designed as twenty-eight closely spaced columns, with gaps to allow airflow straight through.

Figure 57. The Sydney Tower and its model. Courtesy of the Boundary Layer Wind Tunnel Laboratory.

But the designers, Wargon & Chapman, also suggested the use of a damper. The lab examined the details and incorporated the damper into the aeroelastic model used in the test program. At a later stage, however, a number of characteristics of the tower changed. An increase in concrete in the upper turret increased the mass of the structure considerably. And a firewall mandated to separate the elevator and stairwell shafts bisected the column, ruining the cross flow and altering the internal aerodynamics. Construction of the tower was completed finally in 1981, with the damper installed. With all the changes, however, it subsequently became apparent that the tuned mass damper was tuned to the frequency of the tower as designed but not as built—the damper was out of sync with the building. Just before the tower opened, it also became apparent that wind patterns in the now semicircular elevator shaft caused the elevator cables to swirl in larger and larger loops, hitting the wall of the tower. The lab was called in again, and project manager Barry Vickery solved the problem by recommending the installation of glass panels to reduce the gaps between the columns, further preventing the cross flow of air. This, however, brought back the dreaded vortex shedding, causing the tower to have unacceptable vibrations, but this time in the second mode. Ultimately, the first tuned mass damper was never recalibrated. A second tuned mass damper—a pendulum-like mechanism fitted outside the tower, conveniently out of view of tourists—finally put the structure to rest (more or less).

Perhaps the most essential principle of wind tunnel testing, Davenport always argued, is to check the results of wind tunnel model studies with actual observations of full-scale performance once the structure is built—test the test results, so to speak. The CN Tower's program of full-scale measurements began during construction, as soon as the concrete shaft was in place, and lasted for nearly a decade (for another example, see the sidebar, "The Holy Grail of Full-Scale Testing," page 101). The all-hands-on-deck monitoring team in-

Figure 58. With the CN Tower nearing completion in 1975, Artist Rick/Simon designed a three-foot poster, which he mounted on poles in downtown Toronto, within the area dubbed by the media the structure's "fall zone." Invited to climb to the top of the tower by one of the designers, the artist also hung a poster on the interior shaft. "During my visit, which finished by standing on the tip holding onto the lightning rods," he recalls, "I installed a poster, pasting it on the outside of the steel armature but inside the fiberglass tube and covering it with clear Mylar to protect it from the elements." The poster generated a mild media storm, during which the CN Tower's director of design and construction assured the public that 400-miles-per-hour winds would be needed to topple it over. In fact, as the artist discovered when he called the lab to investigate, the structure was tested for sufficiently lower wind speeds—but still, according to the CN Tower report, winds of a substantial 125 miles per hour. For his own part, on visiting the tower for a second time while it was closed during a windstorm, Rick/Simon concluded, "She sways big-time." Courtesy of Rick/Simon.

cluded lab researchers together with the Department of Civil Engineering at the University of Toronto, the Atmospheric Environment Service of Canada, the Division of Building Research of the National Research Council, and Nicolet Chartrand Knoll Engineering, which oversaw the tower's construction. It was an invaluable opportunity, from which all interested parties had much to learn. As noted in a paper assessing the CN Tower's performance on the occasion of its twenty-fifth anniversary, the CN Tower represented "a cutting-edge project of structural engineering practice. As it often happens, sharp edges, such as knives, become jagged when used. They must then be resharpened at the right places. . . . The assessment of structural performance may help towards establishing a list of do's and don'ts for future towers."

The Holy Grail of Full-Scale Testing

Wind engineering's holy grail is to compare wind tunnel results with full-scale tests. Opportunities for this kind of fieldwork, however, are few and far between. Full-scale experiments on large structures are expensive to install and require lengthy monitoring, since the wind is fickle. The majority of full-scale tests measure modest wind speeds because these milder winds occur most frequently. Storm winds are rare by comparison, and gale-force hurricane winds rarer still.

It was extraordinarily fortuitous, then, that in 1982 just such an opportunity presented itself. Hal Iyengar, partner at Skidmore, Owings & Merrill, the designers of the Allied Bank Plaza in Houston, Texas, together with the plaza's owner, Century Development Corporation, agreed to install full-scale testing instrumentation in their new 992-foot tower (owners are often resistant to monitoring owing to the potentially negative connotations: monitoring suggests there might be something wrong, and if tests find there is in fact something wrong or worrisome, this might affect the value of the building and its appeal to clients). The lab rented and installed the equipment at considerable expense, and after several months of testing in meager winds, the project was nearly called off. But then came the news of Hurricane Alicia tracking toward Houston, with winds as high as 115 miles per hour as it neared landfall.

Generally speaking, the primary goal of this full-scale testing was to measure the frequencies of the completed tower (now called Wells Fargo Plaza), since frequencies are an excellent indicator of structural health, serving to confirm advance calculations and validate, Davenport hoped, the wind tunnel study that had been carried out at

Figure 59A. The Allied Bank Plaza in Houston, Texas.
S. Kubyshin/Shutterstock.com.

his lab in 1979. But the wind tunnel study on this project had been far from straightforward. So, as an added bonus, the full-scale tests would provide more nuanced reassurances as well. For one, Houston sits not on bedrock but on a compressible layer of clay. The lab knew from the outset that this was bound to affect the tower's foundation, though rather than jeopardizing the structure's ability to resist wind, investigators suspected the clay would have the paradoxical effect of increasing the inherent structural damping, augmenting the tower's

capacity to absorb and dissipate unwanted motion. The lab's Milos
Novak, an expert on soil-structure interaction, was confident in this
hypothesis, and as a result, the tower's model in the wind tunnel was
tested with more structural damping factored in than usual (1.5 percent rather than 1 percent of critical damping). And in fact, it was
this unique element that motivated the full-scale testing. In agreeing
to the lab's innovative modeling on the damping front, the SOM engineer in charge, Robert Halvorson, insisted on a regime of full-scale
monitoring to ensure that the clay was having the expected cushion-
ing effect. The wind tunnel tests were also challenging because the
tower possessed an unusual geometry, not all square corners, more
curves and indents. Wind engineers are less comfortable with curved
corner geometry as these shapes are more scale dependent and more
temperamental in modeling.

Nonetheless, the lab's study reported extensive information on
wind loads that could be expected to act on the tower's frame during
extreme conditions, such as a severe hurricane, although according
to existing records, tropical storms in the Houston area were a relatively rare occurrence. The structure was designed for a one-in-100-
year wind event, meaning that it was designed for a wind speed that
would on average occur once every 100 years. Such a wind threatened to arrive with Hurricane Alicia on the night of August 18, 1983.

The full-scale experimental equipment was crude by modern standards. The accelerometers and recorders had to be switched on manually at the site. As Alicia swept into Houston, the intrepid Halvorson dutifully went downtown in the wee hours of the morning to
activate the recording. He arrived at the tower as the hurricane
neared its peak. The elevators had been shut down, but a wide-eyed
security guard reluctantly accompanied Halvorson and his colleague,
Mike Fletcher, up the wobbling main shuttle elevator to the sky
lobby, and then, wobbling ever more, up another local elevator to the
top, 71st floor, where the building manager hastily took his leave.
Walking on the top floor was like walking on a large ship in a stormy
sea—the building's natural periods were about seven seconds long, to
rock back and forth. The motion was so pronounced that Halvorson

had to grab whatever he could for bracing as he made his way to the instrumentation. At 4:10 a.m. he and Fletcher succeeded in starting the recording, and then they took the stairs down 71 flights. The recording continued until the analogue reel-to-reel tape ran out, around 5:50 a.m.

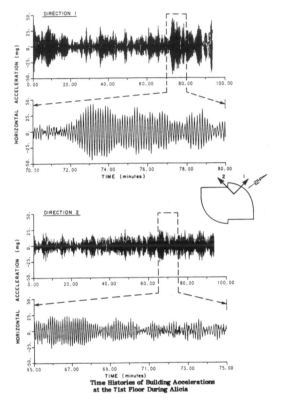

Figure 59B. Printouts from the full-scale measurements taken during Hurricane Alicia. Courtesy of the Boundary Layer Wind Tunnel Laboratory.

During Alicia's peak, the dynamic sway at the top of the building was found to be just under two feet (see figure 59B) The most serious damage sustained by the tower, however, was many broken windows, shattered by ballast rocks blown off the roof of a neighboring building. The full-scale testing instrumentation survived intact and the re-

cording went to Davenport's lab for analysis, which was led by Nick
Isyumov, the project director. First, data were compiled from the
tape to amass a time history of the tower's accelerations as Alicia
blew through town (hitting Houston as well as Galveston directly, the
hurricane caused more than $5 billion in damage in Texas and killed
twenty-one people). Second, using data acquired from Mark Powell
at the Hurricane Research Division of the Atlantic Oceanographic
and Meteorological Laboratory in Miami, the lab's Peter Georgiou
reconstructed the wind field using the lab's newly developed computerized Monte Carlo hurricane model. He reconstructed the wind
speeds and wind directions in downtown Houston during Hurricane
Alicia and used the data to predict the precise winds that hit the top
of the Allied Bank (as it turned out, the wind loads experienced by
the building were about 75 percent of the design wind loadings).

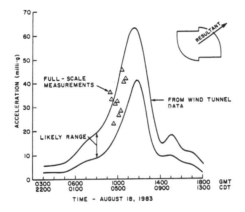

Figure 59C. Graphic comparison of the Allied Bank tower's full-scale versus
model behavior. Courtesy of the Boundary Layer Wind Tunnel Laboratory.

The report from the wind tunnel tests was then used to look up
what had been predicted as the tower's corresponding accelerations.
Finally, the lab compared the full-scale versus modeled behaviors. In
graphing the results, the upper and lower bounds of the modeled response range allowed a ±5-miles-per-hour error band in wind speed
and ±5 degree error in wind direction. Much to everyone's delight,
when the full-scale data were plotted, they fell virtually in the middle

of the error band (see figure 59C, the triangular plots indicating "full-scale measurements"). The wind tunnel tests had been right on the money. The Allied Bank experiment—which remains a rare example of a model versus full-scale comparison with a building put to the test under near design wind loading conditions—bestowed considerable credibility on the new science of wind engineering.

The CN Tower's program for full-scale observations was ambitious. The instrumentation entailed three sets of Bendix-Friez anemometers and directional vanes, mounted on booms at three elevations along the main shaft, with a single anemometer and vane at the very peak. During normal wind conditions, a magnetic tape recorded a continuous 10-minute averaging of data, while during high winds it recorded all data. The sheer quantity of data gathered over years and years made analysis a daunting task, but sifting through the full-scale observations led to one surprising revelation: the earth's boundary layer at the site was "significantly deeper than conventionally assumed." Nonetheless, the tower's actual behavior agreed with wind tunnel predictions. And as the report concluded, "Measuring the periods of vibration is similar to measuring the pulse of an individual. Records to date show a 'steady' pulse and provide a valuable confirmation of the overall integrity of the structure."

An urban landmark and an engineering benchmark, the CN Tower earned its icon status in multiple ways. For the city of Toronto, it signaled the turning point from a backwater to an international metropolis. In 1995, the American Society of Civil Engineers named the tower one of the Seven Wonders of the Modern World, a designation it shares with the Chunnel (the latest addition), the Empire State Building, the Golden Gate Bridge, the Panama Canal, the Itaipu Dam in Paraguay, and the Delta Works in the Netherlands. It draws two million visitors annually, and as tourists and Torontonians alike ascend the concrete shaft in the elevator, they often have a niggling question on their mind: Does the CN Tower move in high winds? The answer is provided in a flier of frequently asked questions. "Like all tall, narrow buildings," it explains, "the CN Tower was built to be flexible and wind resistant." More precisely, in severe winds of 120 miles per hour, with gusts up to 200 miles per hour, the antenna sways about three feet, the SkyPod has about a foot and a half of flex, and the Main Pod deviates an imperceptible nine or so inches from centerline.

In 1976, the year the CN Tower officially opened, the Council on Tall Buildings and Urban Habitat was established at Lehigh University in Bethlehem, Pennsylvania, and published an inaugural quartet of books. One, titled *Tall Building Systems and Concepts*, provided a very readable introductory chapter, a veritable Cook's Tour of tall building structural design vocabulary, which made very clear, as one review noted, that "the dominant influence of the structural configuration of a tall building is our blossoming understanding of wind forces." The late 1960s and early 1970s had been the Wild West of wind engineering—"a bonanza for a handful of experts," as the *Engineering News-Record* described it—with more and more skyscrapers warranting studies. "None of these studies was routine," Davenport assured, in an early report on the lab's first years. With each study the process evolved, incorporating "distinct departures in philosophy and procedure from traditional wind tunnel testing." Davenport also drew attention to the far-reaching benefits of the modern approach. "Information has been developed which was of value not only for the design of the particular structure in question, but also has aided in the piece by piece development of a new and more rational approach to design." This new design approach would serve economic and aesthetic interests, not to mention urgent troubleshooting operations— wind engineering had not quite yet become a tool so common in an engineer's toolkit that all wind-related mishaps were avoided.

On a stormy night in January 1973, with winds at 75 miles per hour, Boston's John Hancock Tower, a mirrored slip of a structure just south of the Charles River, began shedding its windows. From the all-glass façade, entire panes weighing 500 pounds each fractured and fell "like sequins off a dress," as the *Boston Globe* reported in a Pulitzer Prize–winning account of the crisis published two decades hence.

The John Hancock Mutual Life Insurance Company's $95 million skyscraper was still in the final stages of construction when the calamity occurred. Designed by the modernist architect Henry Cobb, of the New York firm Pei Cobb Freed & Partners, "the Hancock" ran to 60 stories, topping out at 790 feet. The curtain wall held 10,344 double-paned windows, and over the ensuing months 3,500 cracked or dropped. Whenever wind picked up speed, authorities closed off surrounding streets. From the sidewalk below, window monitors kept watch with binoculars and doled out window debris to curious tourists. By April, more than an acre of the Hancock was boarded up with plywood and painted black. "Plywood Palace" was one nickname, "Woodpecker Palace" another. A local tourist shop sold T-shirts of a giant woodpecker drilling at the tower, with the caption "Ply in the Sky." The architect Cobb bought T-shirts for his three daughters and one for himself, which he framed.

Speculation as to the cause of the falling windows raised a number of theories. The windows failed because the tower, situated catty-corner to the street, was twisting violently in the wind. The windows failed because the acute angles of the rhomboid structure were creating "hot spots," pockets of high-velocity wind-induced pressure imbalances that sucked off the panes. The windows failed because the building was "soft"—that is, it attracted more wind load than it could cope with. For a scientific diagnosis, MIT engineers hooked up the Hancock like a critically ill patient, with instruments providing full-scale on-site measurement, recording every quiver. The predictions were dire, causing Cobb to send out distress signals to the Swiss engineer Bruno Thürlimann, a leading authority on high-rise steel-frame buildings. Thürlimann in turn commissioned a thorough investigation from Davenport. Davenport's analysis did not agree with MIT's, which had been produced at its Wright Brothers Wind Tunnel—as its name suggests, an aeronautical wind tunnel used primarily for aviation testing. Arguments arose over the quality of Davenport's wind data and his novel use of wind statistics. Ultimately, as Thürlimann's 1975 report stated, "Davenport is an internationally

Figure 60. Boston's notorious John Hancock Tower, featured on the cover of *Engineering News-Record*. Courtesy of *Engineering News-Record*. © The McGraw Hill Companies, Inc., March 27, 1980. All rights reserved.

recognized authority in the new field of wind action on structures." And Davenport's conclusions prevailed.

Prior to construction, Davenport had made a bid to conduct a wind program for the Hancock. Compared to other tenders, however, the lab's analysis was deemed too complicated and too pricey. The architects opted instead to have their design tested in the smooth laminar flow of an old-fashioned aeronautical wind tunnel at Purdue. That the initial tests were not run at the Davenport lab might seem a possible root of the problems. In fact, had the structure been tested in a boundary layer wind tunnel from the get-go, an even more serious wind problem likely would have been detected. While the lab's tests, together with Thürliman's structural analysis, acquitted the wind in the case of the shedding windows (it was eventually discovered that the windows failed as a result of structural flaws in the glass units, but flaws the wind readily exploited), the wind tunnel tests also revealed that the windows were not the Hancock's worst nightmare. The top of the structure was shown to move as much as three feet off center in high winds, putting the tower in real danger of falling over. More unbelievable yet, it was likely to pitch over along its narrow end. When a severe wind excited the tower, its period of vibration went from 12 to 16 seconds. Those extra few seconds would allow the force of gravity to take hold, potentially pulling the tower further and further off from the centerline until it was pulled down, a phenomenon called the P-Delta effect. "It's like moving a ladder around," explains Surry, the lab's lead engineer on the project. "Its own weight helps to tip it over. The generally accepted wisdom within traditional tall building design at the time was that the P-Delta effect could be safely neglected. Since the P-Delta effect is simply the contribution of the displacement of the mass center to the overturning moment, it is not important if the building is relatively stiff and light. For typical compact cross-sectioned buildings, this remains true." With the Hancock, the P-Delta effect was the killer, owing to the tower's extremely elongated cross section.

Conventionally, the wind load on the narrow side of a building was assumed to be proportional to the minimal dimension, with the strength supports calculated accordingly. The Hancock designers followed this reasoning, buttressing the elongated side of the tower with extra support and allowing the shorter side to get away with less. But sometimes the wind is wily and disobeys convention. For the Hancock's design, loads along the structure's long and narrow sides were astonishingly comparable. This was primarily the result of the cross-wind forces generated by the crude airfoil shape, and the vortex shedding. "The design was an efficient and sophisticated one," says Surry. "But less sophisticated designers might have called for the same joints in both directions for simplicity of design, and let the narrow direction be stiffer than they assumed it needed to be." In their sophisticated rush for efficiency, the Hancock engineers undid

Figure 61. The Hancock undergoing much-needed wind tests at the lab with engineer Hiroshi Tanaka, then a research associate. Courtesy of the Boundary Layer Wind Tunnel Laboratory.

their best efforts. And, of course, a truly sophisticated designer would have commissioned a proper wind tunnel test.

Thürlimann presented Davenport's assessment of the situation to his client, detailing "the unfortunate but clear conclusion . . . that the structural performance of the building was inadequate with respect to the safety margins against ultimate load." He recommended a stiffening scheme—strengthening the core with two longitudinal steel frames, and two tuned mass dampers purely to improve occupant comfort. After Davenport ran a thorough analysis of the structure with these modifications, which solved the problem, he told Thürlimann that this study was "possibly one of the most detailed of its kind carried out." That is, until the next calamity came along.

The wind problems of the John Hancock Tower in Boston changed tall building envelope design forevermore. Every design for a skyscraper thereafter received a thorough vetting at a wind tunnel, and for many years the majority of those wind tests took place at Davenport's lab. Davenport traveled constantly, meeting clients and expounding his ideas at conferences, including the importance of the P-Delta effect, even though the Hancock results were confidential. He figured the greater imperative was to get the word out.

Given his international itinerary, and his inclination for rushing here and there, colleagues suggested his name be made a verb—"Davenporting," a variation on "teleporting." He could pull an impressive disappearing act. His office was configured in such as way that there were two doors from which he could exit, one opening into the main office and one into the hallway. His secretary managed to keep track of most of his comings and goings, but she did not realize he had a third exit, which came to light when a graduate student spotted Davenport climbing out his office window with a squash racket under his arm, likely off for a game with then president of the university George Connell, a regular opponent.

Davenport traveled far and wide (faithfully calling home nightly), though he was notorious for always almost missing the plane. Departing from London and inevitably running late, he'd have his secretary call ahead to say he was on his way, and in that bygone era the airline often agreed to hold the plane. Other times he ran out onto the runway in desperation, or left the rental car, keys in ignition, on the airport's departures drop-off driveway. He once boarded a plane in London en route to Chicago and, as was his habit, promptly fell asleep before take-off to catch up on some rest. Awoken by a stewardess, he groggily joined the other passengers deboarding, and began focusing on the day of business ahead. Only after he was off the plane did he discover he had been sleeping mere minutes and, owing to a mechanical difficulty, was still at the airport in London. But usually he departed and returned, departed and returned, without incident. And even after stints of international jet-setting he would arrive home and resume his punishing schedule, depriving himself of the luxury of succumbing to jet lag. He kept tennis dates with his son Andrew at 6 a.m. on Saturdays. He made it to dinner parties that had him nodding off into his soup, and attended guest lectures that sent him into a slumber—though he would rouse himself and rejoin the conversation just in time to make a comment that was insightfully apropos.

The next calamity, as it were, was more just an uncomfortably close call. Nonetheless, it was a near miss that pulled into sharp focus the subtleties of wind analyses and the difference they could make to a structure's fate, literally the difference between a building standing strong or toppling like an ill-set house of cards.

On a temperate day in June 1978, the eminent Boston-based structural engineer William J. LeMessurier received a phone call from a Princeton engineering student who had a question about another corporate trophy tower. At 915 feet, the Citicorp Center was hardly the tallest skyscraper in New York City, though its off-kilter base made it a structural curiosity. The tower's 59 stories sat on four nine-story stilts. And the stilts were not positioned at the usual location, at every corner of the building's square base. Instead they descended from the midpoint of each edge. The Princeton student wondered if LeMessurier hadn't made a mistake.

The quirky stilts were not LeMessurier's mistake with Citicorp, and, as he reassured his interlocutor, this off-kilter scheme was not superfluous architectural flamboyance. The Boston architect Hugh Stubbins had put the pillars where he did to make room for St. Peter's Lutheran Church beneath. The church had occupied the corner of Lexington Avenue and 54th Street since 1905. In 1969 the Lutherans

Figure 62. New York's Citicorp Center was hardly New York's tallest skyscraper, but its off-kilter base, sitting on four nine-story stilts, made it a structural eccentricity. © Norman McGrath, www.normangrath.com.

did some soul-searching, and, as a church history noted, "The congregation unanimously adopted a statement of purpose, which declared, 'We must neither fear nor avoid our mission, but must strive to bring a witness to this city.'" What they witnessed was skyscrapers encroaching from all directions, and Citicorp, the international banking titan, eyeing their property. To secure the church's future, they sold the lot to Citicorp, with one condition. The bank could build its skyscraper as long as the Lutherans got a new stand-alone church at street level—the tower had to cantilever out over the church. When the Lutherans proudly consecrated their new home, they deemed it "a majestic rock, a sanctuary of light, a surprise on Lexington Avenue." The real surprise, however, came a year later, when the structure looming above the church earned the worrisome sobriquet "the tottering tower."

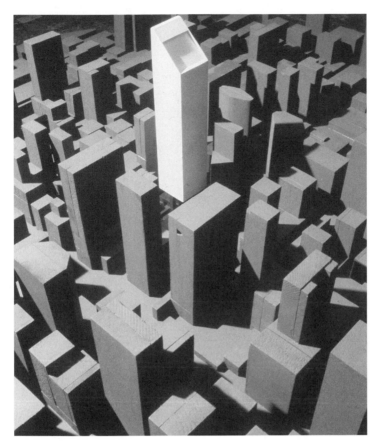

Figure 63. The Citicorp Center underwent a full wind tunnel workup in the planning stages in the early 1970s, an era when wind tunnel tests were still not commonplace for the average skyscraper. Courtesy of the Boundary Layer Wind Tunnel Laboratory.

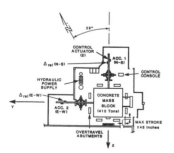

Figure 64. A plan view of the tuned mass damper installed on Citicorp's 63rd floor to counterbalance the wind. Courtesy of the Boundary Layer Wind Tunnel Laboratory.

In 1998 the *New Yorker* ran a feature titled "The Fifty-Nine-Storey Crisis" wherein Davenport made a cameo appearance (for legal reasons, the details of the tower's travails were kept secret for nearly two decades). The lab had done a full workup on the Citicorp Center in the planning stages in the early 1970s. A 1974 article in the *New York Times*, "The Wind, Fickle and Shifty, Tests Builders," had quoted Davenport commenting on the project and on the advisability of wind tunnel testing in general. "Dr. Davenport said that wind speeds of more than 35 miles an hour are not uncommon at ground level in urban areas. Such winds, he added, are dangerous as well as unpleasant, for they create situations in which the pedestrian finds it difficult to maneuver."

That the Citicorp tower underwent testing was unusual for the time. It was prior to the Hancock crisis, in an era when wind tunnel tests were still not commonplace for the average skyscraper. In that day, wind tests were prescribed for one of two reasons: if the structure, in its architectural daring, could do with some extra insurance, or if the structural engineer was forward-thinking enough to realize the benefits a wind tunnel test offered in optimizing design. Otherwise, wind tunnel tests were executed when there was a problem, after the fact. Initially, LeMessurier's decision in favor of wind tests fell within both categories one and two. Proactively, the tower's original plans even included two tuned mass dampers—two 400-ton slabs of concrete that lay on oil-film bearings, with a computerized actuator programmed to counteract sway as necessary. Only one was installed, reducing the building's sway amplitude by 50 percent. The cost of the damper system ran to $1.5 million, but in return, LeMessurier estimated it would save $4 million—the cost of a few tons of structural steel that otherwise would have been needed to stiffen the frame in making the tower's motion less bothersome to occupants. In this respect, Citicorp was a textbook example of good wind engineering practice. And as the *New Yorker* feature noted, it was an example LeMessurier himself showcased in teaching his engineering students:

> A classroom lecture would also look at the tower's unusual system of wind braces, which LeMessurier had first sketched out, in a burst of almost ecstatic invention, on a napkin in a Greek restaurant in Cambridge:

forty-eight braces, in six tiers of eight, arrayed like giant chevrons behind the building's curtain of aluminum and glass. ("I'm very vain," LeMessurier said. "I would have liked my stuff to be expressed on the outside of the building, but Stubbins wouldn't have it. In the end, I told myself I didn't give a damn—the structure was there, it'd be seen by God.")

The unique geometry of the bracing system was designed to route all the wind-induced overturning forces down to the pillars at the center points, and from there mounting the best defense. The braces were designed with welded joints, fusing together two structural steel members to make them as strong as one. LeMessurier explained all this to his Princeton interlocutor over the phone, and afterward, just for the heck of it, he refamiliarized himself with some calculations he'd done to prove the braces' brute strength.

According to the New York City Building Code, the braces had to be strong enough to withstand perpendicular winds, hitting each face of the tower head-on. As LeMessurier tells it, his replay of the calculations confirmed that the joints more than met this standard. Then, feeling intellectually playful, he decided to do the calculations on the strength of the joints in quartering winds—winds hitting the structure on a 45-degree diagonal, thus stressing two sides at once and putting more stress on each. These calculations took him by surprise. With four of the eight chevrons on each tier, the quartering winds increased the load by 40 percent. The welded joints would have absorbed this extra load. But, he recalled, a last-minute decision during construction had replaced the welded joints with bolted joints—the more expensive welded joints were deemed stronger than necessary, and thus an unnecessary expenditure. When the decision was made for bolted joints, the effect of quartering winds had been overlooked. LeMessurier's once overweening confidence in his system of wind braces was shaken.

For his own erudition, LeMessurier detailed these mistakes, and a few others that unfolded in a chain reaction of miscalculations and missteps, in a confidential document he wishfully titled "Project SERENE"—"Special Engineering Review of Events Nobody Envisioned." He then turned to Davenport's team of wind engineers. He wanted them to revisit the original evaluation in the light of the new information about the bolts. "If we are going to think about such things as the possibility of failure," LeMessurier noted, looking back, "we would think about it in terms of the best knowledge that the state of the art can produce, which is what these guys could provide for me."

LeMessurier flew to London, broke the news, and insisted on the truth straight up. After the lab's analysis, LeMessurier met again with Davenport, this time in New York. The stress of the crisis was compounded by Davenport's pesky travel habits. He was traveling from overseas, but he was without his Canadian passport (he'd forgotten it at home), thus making his entrance into the United States for this urgent meeting far from certain. Isyumov was traveling to the meeting from home base at the lab with Davenport's passport in hand, making for some tense moments in coordinating the "passport transfer" at the airport.

Davenport made it to the meeting and proceeded to detail the lab's conclusions, which were even more alarming than LeMessurier's estimates. Taking into account the intervening five years of available meteorological data, the theoretical increases LeMessurier had scratched out would increase even more, especially in hurricane-scale winds. And since hurricane season was approaching, the lab provided LeMessurier with a lookup table, a crib sheet of sorts, laying out what storms of varying severities would do to the tower, which ranged from damage to destroy.

Armed with this analysis, LeMessurier broke down the stresses from the wind forces for each and every structural member of the building and determined that the weakest link was the joint on floor 30. If that joint gave way, so would the building. He also determined that the probability of a storm capable of delivering such a blow was once in every sixteen years, a "sixteen-year storm"—or one chance in sixteen in any given year. Statistically, the tower was vulnerable. In lay terms, it was in grave and imminent danger.

The engineer considered his options. He could remain silent. He could commit suicide. Then, as recounted in the *New Yorker*, he had an epiphany. "I had information that nobody else in the world had. I had power in my hands to effect extraordinary events that only I could initiate. I mean, sixteen years to failure—that was very simple, very clear-cut. I almost said, 'Thank you Lord, for making this problem so sharply defined that there's no choice to make.'" A series of tense meetings ensued: with the architect, with the insurance company, with liability lawyers, and with Citicorp executives. The Red Cross was called in to devise an evacuation plan, should the sixteen-year storm hit. Round-the-clock technicians made sure the damper system was in working order, with emergency backup generators at the ready—the damper had been installed for creature comfort, to make any sway imperceptible; now it was the life preserver, increasing the return period of failure. And structural engineer Les Robertson was called in to oversee the repairs. The emergency action plan was to weld two-inch-thick steel plates over each of 200 bolted joints. Since the tower was occupied, welders worked beneath plywood shacks erected in the middle of tenants' offices, Monday through Sunday, but only at night.

Day by day, as the welders progressed joint by joint, the lab made running calculations diagnosing which floors held the next greatest risk and dictating which joints should move to the front of the welding queue. By the beginning of September, with Hurricane Ella and its 140 mile-an-hour winds heading for New York, evacuation plans were stepped up, though the building at that point had been fortified to endure a 200-year storm. And then Ella turned course, heading toward Nova Scotia. By October, just in time for the end of hurricane season, the rehab was complete. The formerly tottering tower was now fit to withstand a 700-year storm.

There is a coda to the mythologized Citicorp tale of the quartering winds. LeMessurier's *New Yorker* confession implied that the quartering winds were not explicitly factored in at the design stage. In fact, the lab's findings had emphasized that both the perpendicular and the quartering winds should be taken into consideration, acting individually and simultaneously. This was a key part of the lab's innovative approach: it tested wind approaching from every 10 degrees around the compass and synthesized all directions to determine the most critical three-dimensional loading scenarios. It was up to the structural engineers to take heed. Alas, traditional engineering practice and the building codes held that it was only necessary to consider wind loads from one direction at a time. "Bill had not taken into account the combined action of wind-induced forces acting in both principal directions of the building," says Isyumov, who worked on both the initial wind tests and the emergency follow-up. "The need to consider this joint action was clearly reported in our wind tunnel study. His structural system was equally sensitive to both sets of these forces and was therefore overloaded. His subsequent explanation of attributing the problem to the action of 'quartering winds' was not correct. In fact, the maximum wind forces on his building were due to approach winds parallel to the sides of his building and definitely not those which came from a 45 degree direction. Neither Alan nor I took issue with Bill's 'official' explanation, which involved some 'face-saving.' He was one of our loyal clients, and after all he did take the necessary corrective measures."

Better late than never. Educating engineers on the merits of abandoning their old-fashioned perspective required diligent discussions and explanations of wind reports, and even then sometimes to no avail. In the chaos of last-minute decisions, subtleties could easily be lost or overlooked. Wind-smart structural engineers gradually became more the norm than the exception, with codes evolving in tandem, or lagging a little behind owing to the complex committee processes that had to take place before new codes could be published. The National Building Code of Canada first stipulated that the combined action of wind loads should be taken into account in its 1980 edition (Davenport, who had been privy to the otherwise proprietary information detailing the Hancock crisis caused by unbalanced torsional loads, sat on the executive committee). The American code, set by the American Society of Civil Engineers in a tome titled *Minimum Design Loads for Buildings and other Structures*, didn't introduce the new standard until the 1995 edition.

Codes also steered a conservative course for the Hongkong and Shanghai Banking Corporation's new home, completed in 1986. All blueprints and designs had to meet the approval of the local Building Ordinance Office, which often exercised its veto power. Architects usually obliged and created buildings in pre-approved styles. For the Hongkong and Shanghai Bank, however, all candidates considered for the job were known for their distinctive structures. The job went to Britain's Sir Norman Foster, who at the time had not completed a structure above three stories tall.

"The brief for the Hongkong and Shanghai Bank Headquarters was a statement of confidence: to create the best bank building in the world," noted the firm Foster and Partners. "Through a process of questioning and challenging—including the involvement of a feng shui geomancer—the project addressed the nature of banking in Hong Kong and how it should be expressed in built form. In doing so it virtually reinvented the office tower." Constructed in Foster's self-described "High Tech Modern" style, the 590-foot tower, all steel and glass, showed off the architect's innovative "coat hanger" structural scheme on its exterior, with an interior likened to a *Star Wars* set (Foster was known to find inspiration in the aircraft and aerospace industries, and this project strove to meet perfectionist standards usually reserved for nuclear or defense construction). Like the Citicorp Center, this tower did not rest on the ground. It hovered on steel columns, since the new bank occupied the exact site of the old bank and the owners stipulated that the original building remain in place, and fully operational, during construction. It would be demolished only after the tower's completion, to clear space for an open plaza. Tearing down an old icon before a new icon was in place was thought to imperil Hong Kong's prosperity.

The structure had a modular design, with five main units that were prefabricated and assembled by robots (rumor had it the modular design was just in case the tower had to be disassembled and moved, should there be any problems with the 1997 handover of Hong Kong from Britain to China). The bottom line of all these innovative eccentricities earned the tower the reputation of being the most expensive structure in the world, with a reported price tag of U.S. $1,000 per square foot, for a grand total of $800 million.

It is hardly surprising, then, that the wind study was as high tech as the structure undergoing tests. The threat of typhoons meant the climate studies were more exhaustive than ever. And the open plaza beneath meant an experimental study of the wind environment effects on pedestrians was necessary. The lab had ventured into these parts with a 1975 study of Commerce Court, a cluster of office towers in downtown Toronto's financial district so impassable on a windy day that rope lifelines were installed to assist pedestrians as they approached. And in 1978 it had studied equally treacherous terrain around the American Telephone and Telegraph Company and IBM corporate headquarters in New York. That study provided pioneering verification that wind tunnel tests could be used effectively to predict pedestrian-level wind, with Davenport taking the lead by suggesting standards of "comfort criteria." For the Hongkong bank tower the lab proposed various "aerodynamic amelioration techniques," including canopies, patterns of roughness elements to break up the flow, and partial closure of the north or south open walls. The latter was the best option but unacceptable since the space had to remain entirely open. Partial glass walls provided the solution, hanging from the third level to within about 10 feet of the ground.

Figure 65. The Hongkong and Shanghai Bank model laid out on the university's Alumni Hall stage for filming. Courtesy of the Boundary Layer Wind Tunnel Laboratory.

Figure 66. The model for the Hongkong and Shanghai Bank, like the tower itself designed by British architect Sir Norman Foster, spared no expense. Courtesy of the Boundary Layer Wind Tunnel Laboratory.

However, the most pressing factor on the wind tests was time. First, there was Hong Kong's strict custom that construction occur over the shortest possible duration. Second, Foster himself relished beating a deadline, and he liked to push the process to the limits, incorporating new modeling and testing phases even when the schedule was already tight. At Davenport's lab, the telescoped time-table kept engineers working into the wee hours of the morning, occasionally with flights the next day to England for meetings. And so there could be no better opportunity to deploy a novel modeling technique that the lab had in the wings—a technique that would save precious time.

The "base-balance" model, devised by Tony Tschanz, then a PhD student at the lab, depended more on the technology of the balance at the base than on the modeled structure that stood above. Later to be called the "force-balance" model, an ultrasensitive five-component balance at the base measured the forces exerted on a simple structural model cut from rigid but lightweight foam. Whereas an aero-elastic model could take months to produce, the new base-balance model took three weeks. Looking at the big picture, this was a time frame much more in tune with the demands of the industry, making the commercialization of wind studies more feasible.

Nevertheless, all facets of the wind study on the Hongkong and Shanghai Bank took two years, from 1980 to 1982, consuming 1,000 man-hours just to make the pressure model. The client demanded that all the architectural details be replicated. The lab assured this wasn't necessary, but the client insisted. The bill for the pressure model ran to the mid-five figures. The lab produced eleven different reports. The protracted timeline notwithstanding, the client was apparently pleased with the job, as it gave the lab a gift of $800,000 toward the construction of a new wind tunnel in 1985 (though the gift was also in recognition of the large numbers of Hong Kong students studying engineering at the university).

The lab's wind study concluded that Hong Kong's codes for wind loading followed the trend for conservatism, and that the bank, sheltered by the 1,811-foot mountain Victoria Peak, was in good standing with the wind. The tests also revealed that, with the exception of the seasonal typhoons, Hong Kong's day-to-day wind climate was significantly calmer than temperate regions'—Davenport remarked in a letter to Foster that Hong Kong winds had only 75 percent the strength of winds seen in a city like London. As a result, as Foster noted, "It ought to have been possible to make some savings in the steel used in the frame; unfortunately, the tightness of the pro-gramme and the need to release information for construction did not allow time to negotiate a reduction in the design wind load for the building with the Building Ordinance Office." The conservative codes again won out. But the bank tower would surely outlive them, with its structural frame having an estimated life span of 10,000 years.

To this day, it can take great powers of persuasion to clear the vestiges of outdated wind engineering ideology from the codes. One of the lab's recent projects, for instance, was the Shanghai World Financial Center. Voted best tall building in 2008, it was designed by the New York firm of Kohn Pedersen Fox, with Les Robertson again as the structural engineer. More than a decade in the making, at 1,614 feet it is currently the world's third tallest completed building to its rooftop. The main wind quandary with this structure was more political than scientific. At issue was the quirky aperture planned for the tower's thin square peak. The original design called for a circular hole, in part a reference to Chinese mythology (a circle being the symbol for the sky and the framing square the symbol for the earth), and in part out of deference to the shape with the lowest-resistance and lowest-stress aerodynamics. However, the mayor of Shanghai protested: the circle framed by a square looked too much like the rising sun on the Japanese flag. The architects and the engineers settled on a trapezoidal aperture instead.

Figures 67A–B. Shanghai Financial Center with the originally proposed circular aperture at the peak, and the finished structure with its trapezoidal aperture, which is said to resemble the world's largest beer opener. (A) Courtesy of Leslie Robertson. (B) Cuiphoto/Shutterstock.com

A more substantive issue, addressed at length during one of Davenport's visits to Shanghai, was China's building code and its excessively conservative wind design speeds, still on the books from a bygone era when wind engineering was at best a fuzzy science. After endless meetings (which Davenport did not attend, there being little point) and continuing debate, Robertson succeeded in getting the design wind speeds reduced by 20 percent, and even then they were more conservative than they needed to be, causing undue expense on a project that suffered from a shortage of funds during the Asian financial crisis.

The otherwise smooth sailing that characterized the Shanghai tower is testament to wind engineering's current maturity. No longer is every model and every test an adventure into the unknown. Even the Burj Khalifa in the United Arab Emirates, the tallest man-made structure ever built, made for a relatively straightforward wind testing process, if exhaustive. The Burj overtook the CN Tower as the world's tallest freestanding structure in the middle of construction in September 2007. The tower opened in January 2010, its pinnacle, the height of which was long a closely guarded secret, topping out at 2,717 feet. At those altitudes, the wind in some ways was a major enemy. "It was essentially designed inside a wind tunnel," says John Zils at SOM, the project's architectural and engineering firm. "We now use the wind tunnel as a design tool, not just as a technical tool that confirms design." Davenport's instinct for the role of wind analysis, and his intuition for architectural and engineering issues, had met their natural union. On a practical level, it meant that both the structural engineer and the architect were often present at wind tunnel tests—still multiple tests, iteration after iteration, but with changes made on the spot as the design was sculpted to perfection by the wind.

The Burj wind tests were conducted at the firm of Rowan Williams Davies and Irwin, in Guelph, Ontario, where many of Davenport's descendants have landed over the years. A second opinion, or peer review, was conducted by the Davenport lab in the form of a parallel wind tunnel test. This extensive testing on the Burj revealed a peculiar phenomenon: when the tower reached a certain (secret) height, the wind forces became smaller, not bigger. But only after wind tests probed every nook and cranny of all the micro-design elements did

Figure 68. For the Burj Khalifa in the United Arab Emirates, the tallest structure ever built, the Davenport lab conducted a peer review of the original wind tests performed at the firm of Rowan Williams Davies and Irwin, in Guelph, Ontario. Rahhal/Shutterstock.com.

this strange condition assist the structure in its ascent. The tower's spiral was reversed, from counterclockwise to clockwise, to align with the dominant wind direction and funnel the wind forces to the structure's advantage. And the window mullions, likened to the skin of a shark, were crafted to create optimal eddies, lubricating the wind as it passed over the building and minimizing its effect. Only with engineers tending to these microscopically fine details in the wind tunnel could the Burj grow so tall.

Even with the engineering science holding it up, the heights of the Burj become nerve-wracking for engineers watching from near and far. "We may cross a tipping point without knowing it," says RWDI chairman Peter Irwin. Irwin first met Davenport in 1977 as a young engineer with Ottawa's National Research Council, where Davenport made a visit to observe aeroelastic model tests of the Lions Gate Bridge. The fledgling Irwin was nervous at the prospect of such a towering and illustrious figure observing his tests, but he need not have been. As ever, Davenport, a man known for his many kindnesses and his modesty, put everyone at ease with generous praise, positive encouragement, and advice for senior and junior engineers alike. He was determined that all wind engineers move forward together.

"Many of the methods of modern wind engineering owe their origins to him," Irwin says. And it is on that foundation that the field continues to advance. An even taller supertower is already in the works in Dubai—more than doubling the Burj, venturing upward 1.5 miles and 400 stories, beyond the earth's boundary layer and into the free atmosphere. If at these heights a tipping point is ever reached, Irwin notes, the nature of the testing might have to change. But as yet, techniques developed by Davenport nearly a half century ago are still the standard, though theoretically refined and technologically advanced by orders of magnitude. The 2,717-foot Burj followed the template set out by the seminal, though shorter by half, World Trade Center. The World Trade Center was the step change, the proverbial paradigm shift or quantum leap that allowed a credible scientific evaluation of the stochastic forces of the wind. And so while most of the hundreds of models tested at Davenport's laboratory reside in the backrooms, stored helter-skelter in an international metropolis of the tallest buildings and longest bridges, those seminal twin towers, as well as other select structures (and a kite), still greet visitors in the lobby not ten steps from the entrance, directly in the path of the occasional gust of wind rushing indoors.

III
Long and Longer Bridges

During a sabbatical spent in Belgium and France, the Davenports took the kids out of school and, for a time, set up camp in a trailer park. During a weeklong visit to Paris they parked their trailer in the Bois de Boulogne at the edge of the 16th arrondissement. Davenport emerged daily from these digs to find a driver waiting to chauffeur him to a few lectures here and there. Predominantly, however, he resided in an attractive house on avenue Hamoir in Brussels, where he consulted on the construction of a new wind tunnel at the NATO-sponsored von Kármán Institute for Fluid Dynamics. For the weekends, the family set off on road trips across Europe, with the kids asking the typical questions as to where they were going and how long till they got there. The little Davenports soon learned that after driving for hours these outings always ended the same way: at one bridge or another.

Bridges were his first structural love, yet Davenport was catholic in his wide-ranging interests (philosophically speaking he was agnostic, despite what he considered an excess of religious teachings at Anglican boys schools). A Renaissance man of sorts, he cultivated a passion for politics, poetry, art, classical music and the blues, theater and cinema (his favorite film: *The Gods Must Be Crazy*). And all of the kids, Tom, Anna, Andrew, and Clare, each have a quintessential memory of Dad's at times ad hoc engineering genius as distinct from the inevitable destination of those road trips. For instance, he crafted a backyard tennis backboard from plywood and a foam mattress so as to precisely control the ball's rebound speed. He constructed handmade parallel bars and a balance beam that fit into the living room. His makeshift briefcase holder, nicknamed the "nerd box," was attached to his bicycle with bungee cords and adorned with orange cautionary flags. He relented at his children's pleading and delivered on his promise of a color television one Christmas, retooling a black-and-white television by painting its exterior that same cautionary orange (he had an acute, often droll, sense of humor, though he rarely laughed aloud, and when he did, not loudly; he would just look very amused). He took great offense when anything broke, and started in immediately on the repairs. And he always kept at hand a pad of graph paper, as well as a .7mm mechanical pencil, his media of choice for scribbling out not only professional matters but also a Christmas poem for Sheila, or a note left on the dining table indicating he'd gone to the university for a minute, a minute that invariably morphed into hours. Even at home, when seized by an idea or a challenging engineering problem, he could plunk down where the noise and chaos were worst and block it all out, intoxicated by intellectual tinkering. Once Sheila entrusted her husband to take care of their four young children for an afternoon, only to return home to find

Figure 69 (left). Davenport atop the tower of the Severn Road Bridge, crossing the River Severn joining England and Wales, circa 1966. Courtesy of Sheila Davenport.
Figure 70 (right). During their father's sabbatical in Belgium and France in 1972, the Davenport children came to enjoy visiting many civil engineering structures, especially bridges (though here they seem to be posing on a canal lock). Courtesy of Sheila Davenport.

the kids crazed, tearing apart the house. As the eldest child, Tom, recalls, Dad was present at the center of this whirlwind, intently focused and intensely alive, sitting in the middle of the playpen, working away. He had an ability to concentrate in the eye of any storm.

Above all, however, Dad's defining characteristic was his love of bridges. As his daughter Anna recalls, the Davenport progeny all learned to identify a suspension bridge at a very early age. He talked of bridges even more than he did of wind, though the kids were well versed in wind as well; Andrew executed a sophisticated grade school science project on windmills. But bridges embodied his true and deep obsession. He documented the year in Europe with rolls upon rolls of photographs of bridges, though comparatively few of the family—nary a shot capturing the frequent picnics *under* bridges or *beside* bridges while Dad extolled the virtues of the marvelous structure.

ᴝ   ᴝ   ᴝ

The beauty of a bridge, Davenport would argue, is there for any observer to behold. More so than a building since a bridge's engineering is exposed, naked to evaluation—the structure should indicate where the loads are going and how they get there. Suspension bridges, for example, rely on cables to carry the load from the road deck to the towers and anchorages. At the anchorages the cables take root, depositing the horizontal load into the earth, and at the towers the vertical load is transferred into the pier foundations. This structural performance should be easily apparent, not masked or falsified—a beautiful bridge is an honest bridge. Honest or not, there was rarely a bridge that Davenport did not appreciate.

A procession of nearly one hundred bridges traveled through Davenport's lab, and some, though certainly not dishonest, were imaginatively misleading in their structural creativity. The bridges by Santiago Calatrava, the architect who doubles on the artistic axis as a sculptor and on the scientific axis as an engineer, are paradoxical structures, static embodiments of dynamic motion. Technical treatises and exhibit catalogues on Calatrava bear titles such as *The Poetics of Movement* and *Structures in Motion*. His Alamillo Bridge in Seville, Spain, is off-kilter in its asymmetry, with only one 465-foot-high pylon reclining an obtuse 122 degrees from the roadbed, its leaning

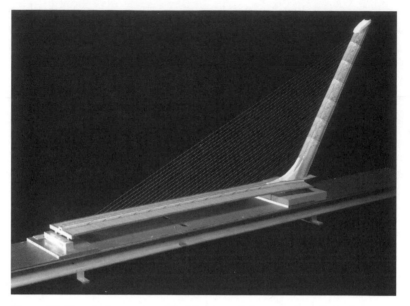

Figure 71. The wind tunnel model of Santiago Calatrava's Alamillo Bridge in Seville, Spain, an engineering work of art, now resides in the Pompidou Centre in Paris. Courtesy of the Boundary Layer Wind Tunnel Laboratory.

mass apparently offsetting that of the bridge and the traffic. Resembling a giant boomerang, it looks more apt to lift and fly than sit steady and stationary. Wind tunnel tests scuttled plans for a public observation platform within the upper levels of the solitary pylon, but otherwise there were no worrisome wind effects. As expected: Calatrava had insisted at the outset that the bridge would be sufficiently stiff, that its high natural frequency would put it out of the wind's reach.

Most bridges, however, are not beyond the reach of the wind. One bridge in particular, New York's Bronx-Whitestone Bridge, built between the world wars and opened in 1939, has made numerous visits to the lab over the years for wind-related recalibrations. With this provenance it serves as a nice touchstone, a chronicle of how the aerodynamics of bridge engineering has evolved over the course of a century.

Crossing the East River, linking the Bronx with Queens on Long Island, the Whitestone, as it is known, has a main span of 2,300 feet (the main span being the standard measure). Designed by two eminent bridge engineers, Leon S. Moisseiff and Othmar Ammann, it opened as the fourth-longest suspension bridge in the world just in time for the 1939 New York World's Fair—the fair, in fact, was the driving force behind its construction and hasty completion in less than two years. At the opening ceremony, Robert Moses, then chairman of the Triborough Bridge Authority, declared it "architecturally the finest suspension bridge of them all, without comparison in cleanliness and simplicity of design, in lightness and the absence of pretentious ornament. Here, if anywhere, we have pure, functional architecture." That was the architectural, aesthetic opinion. Ammann articulated the engineering perspective: "The new structure breaks no records for length of span, but it embodies a number of departures in design that emphasize, more strongly perhaps than mere increase in size, the rapid progress that is taking place in the bridge-builder's art."

Not two years later, as Davenport so often liked to recall, the Tacoma Narrows Bridge earned its infamy, felled by a 39-mile-an-hour wind—a moderate to fresh gale, according to the Beaufort scale, a fairly common wind capable of snapping twigs off trees and impeding progress if one were caught out for a walk. Lauded as the latest in bridge technology, this bridge had also been engineered by Moisseiff, known for his deflection theory of suspension bridges—a theory that pushed the rapid progress in the art of bridge building, liberating engineers to design specimens boasting ever-increasing physical flexibility and slenderness. Technically expressed with a string of formulas, though simple enough to offer back-of-the-envelope cachet, Moisseiff's theory proposed that the deflections of the deck and the cables worked in tandem. Setting it apart from previous theories, the deflection theory factored in the actual deflection of the cables and deck in the stress analysis, in turn facilitating the design of more flexible structures. As deck stiffness decreased, the suspended structure's dead weight took up the slack in limiting deflections. And so as bridge spans lengthened, and suspended weight increased, deck systems became more and more flexible. Moisseiff's formulas calculated the behavior of the deck and cables by taking into account the dead loads of the suspended structure itself and the live loads of traffic. But to its detriment, the theory omitted the aerodynamic variable, the live loads caused by wind. Aerodynamics were barely on engineers' radar in the early twentieth century. This was not due to any lack of precedent but rather to an unfortunate blind spot for history.

One hundred years prior to the Tacoma bridge's collapse, much debate had taken place over whether suspension bridges were safe under any circumstances. There were known wind problems, and the early to mid-nineteenth century had witnessed a slew of failures. In the day, John Roebling, a German American civil engineer, studied these failures and devised three basic principles to shepherd the construction of safe bridges. First, a bridge must be of sufficient mass and inertia to quell excessive wind excitation. Second, since the wind can either lift up or push down on the bridge, the structure must have stays tying down its deck, either deck to ground or tower to deck. And third, to be adequately stiff, the bridge must have trusses. Roebling built a number of bridges with this failsafe formula, most notably the Brooklyn Bridge, completed in 1883. "He had a recipe and his bridges were successful," says Henry Petroski. Perhaps too successful. The Brooklyn Bridge emboldened American engineers to build suspension bridges more brazenly. And with their bravado they gradually cut away at Roebling's recipe, one ingredient at a time.

First went the cable stays, when engineers decided they were redundant with the main suspension cable. "What drove the changes principally was a belief that there was a greater understanding in how to analyze bridges, not with models but with mathematics," says Petroski. The mathematics, however, did not include the aerodynamic factors that Roebling had intuited. And by the 1930s, the truss was eliminated, the presidential (and precedential) example being the George Washington Bridge over the Hudson River, designed by Othmar Ammann, with a main span almost twice that of any previous suspension bridge. And although the George Washington later received its second deck, which served to increase the bridge's mass and stiffness, when initially constructed in Depression times it was afforded only one upper deck, producing an exceptionally slender profile and making lithe and lean the trendy new aesthetic.

Then, in the late 1930s, suspension bridges migrated from urban settings to more rural locales—such as Tacoma, Washington. In these parts bridges weren't expected to carry a lot of traffic, so they could be designed with a mere two lanes. "They were narrow, hence they were light. They didn't have a truss so they were shallow and slender. And they didn't have cable stays. So all three of Roebling's components for a successful suspension bridge had been eliminated," says

Figure 72. The George Washington Bridge's slender profile made lithe and lean the trendy new aesthetic in bridge design. Liz Van Steenburgh/Shutterstock.com.

Petroski. "Everybody knew the wind blew. But the way bridge engineers were handling the wind in the 1930s was to say, 'The wind is going to push the bridge sideways, the deck between the towers is going to be pushed sideways. So we have to worry that the towers are stiff enough to stand up to the wind, just as if they were skyscrapers.' They worried about stiffening the bridge in that direction, the transverse. They didn't worry about stiffening it vertically." Then came the Tacoma Narrows disaster, shaking the profession to its foundation and returning bridge engineers to an era of skepticism and conservatism. From the ruin emerged a long overdue science of bridge aerodynamics.

The Tacoma Narrows Bridge failure focused attention on its sister bridge, the Bronx-Whitestone Bridge. The Whitestone oscillated in the wind even before it opened in 1939, exhibiting a similar if less violent vertical vibration. Following the Tacoma disaster, the Triborough Bridge Authority installed a series of stiffening stay cables on the Whitestone that extended from the tops of the towers to the deck's plate girders at the quarter points of the main span. Moses commented that this remedial measure "was like lifting a man up by his suspenders," and just about as effective. Then in 1946, after more complaints from drivers, the bridge received further fortification with the installation of two 13-foot Warren trusses (and in turn there came complaints about the trusses obstructing the Whitestone's spectacular sightlines and obscuring its svelte silhouette). All the while Ammann, ultimately responsible for the bridge's design, insisted his structure was solid and safe. As Moses recounted, "Ammann kept saying in his best Swiss brogue, 'The britch is safe, the britch is safe,' and we kept saying, 'That doesn't make a damned bit of difference if drivers won't use it.'"

Having served on the board of engineers established by the Federal Works Agency to conduct an autopsy on the Tacoma bridge's demise, Ammann was well aware of what was at stake, but also of the difficulties in pinpointing wind issues. Submitting its report in March 1941, the board concluded that the failure occurred when a midspan diagonal stay, added post-construction between the cable and the stiffening girder to give the bridge some help maintaining its composure, slipped out of place. This allowed the structure's relatively benign vertical oscillation, which occurred in rolling waves along the length of the roadway, to devolve into a more malevolent torsional twisting along its width. To make matters worse, the torsional oscillation was of an antisymmetrical mode—at a node, or point of no motion at the middle of the bridge's length, the deck twisted in opposite directions on either side. The error on the part of the designers had been in devising "an extraordinarily small width of structure as compared to its span. Their expectation that the bridge would have adequate lateral rigidity under wind pressure was unquestionably justified, but the combination of this narrow width with great vertical flexibility, proved to be responsible for the excessive torsion under aerodynamic action which led to the failure of the [deck]. This experience reveals the danger of excessive narrowness." The bottom line, as the board advised, was that aerodynamics must become a pivotal factor in bridge design forthwith.

There was no consensus, however, as to the exact aerodynamic role the wind played in exciting the Tacoma Narrows Bridge. The wind

Figure 73A. The Bronx-Whitestone Bridge with its slender configuration opened in 1939. Courtesy of the MTA Bridges and Tunnels Special Archive.

Figures 73B–C. The spartan configuration of the original Whitestone Bridge (left) proved susceptible to instabilities similar to those experienced by the Tacoma Narrows Bridge. By 1946, diagonal stays braced the cables, and triangular trusses ran along the bridge deck (right). Courtesy of the MTA Bridges and Tunnels Special Archive.

tunnel results were inconclusive, perhaps not surprisingly, since these tests subjected the bridge exclusively to smooth laminar flow. The investigators each harbored their own theories—random buffeting, flutter, vortex shedding, some combination thereof, or one leading to the next to the next. The structural engineers remained unconvinced by the aerodynamicists, and vice versa. Board member Theodore von Kármán, then director of Caltech's Guggenheim Aeronautical Laboratory, where wind tunnel tests on a nearly 20-foot-long full model of the doomed bridge were conducted as part of the post mortem, argued that flutter was key. And he criticized his colleagues for the depth and endurance of their prejudices. "Their thinking was still largely influenced by consideration of static forces," he later commented, "like weight and pressure which create no motion, instead of dynamic forces, which produce motion or change of motion. . . . Bridge engineers, excellent though they were, couldn't see how a science applied to a small unstable thing like an airplane wing could also be applied to a huge, solid, nonflying structure like a bridge." Another prominent engineer, David B. Steinman, whose omission from the board was conspicuous, was particularly dissatisfied with the findings. He engaged in a rancorous and public debate with Ammann and pilloried the report's shortcomings in a letter to the *Engineering News-Record*: "[The report] leaves many questions unanswered. It does not tell what combinations of cross-sections produce aerodynamic instability, how aerodynamic instability can be reasonably predicted or readily tested, nor how it can be prevented. There are ways of designing sections to avoid or eliminate aerodynamic instability, and that is perhaps better engineering than merely placing all of the emphasis on adding rigidity to provide increased resistance to a dangerous inherent characteristic. It is more scientific to eliminate the cause than to build up the structure to resist the effect."

For Davenport, these sentiments proved prophetic. Predicting and preventing the aerodynamic instability of bridges would become a going concern for structural engineers for the balance of the twentieth century and beyond.

At the historic 1966 International Symposium on Suspension Bridges in Lisbon, exceptionally well attended by 177 engineers from around

Figure 74. In 1960, the Severn Road Bridge, then in the design stage, became the first aerodynamically streamlined suspension bridge, undergoing tests at England's National Physical Laboratory. Bill Brown of Freeman Fox and the NPL's Kit Scruton engineered the cross section to avoid instability in the design wind speed range, thus halving the weight of the bridge compared to post-Tacoma designs. Although not employing a boundary layer wind tunnel, the NPL used section models and introduced turbulence using grids in aerodynamic wind tunnels (though as Davenport liked to recount with amused chagrin, the NPL had recently demolished a wind tunnel that could have plausibly doubled as a boundary layer tunnel in order to make room for a new cafeteria). Ian Britton/Freefoto.com.

the world, Davenport was again preceded by his reputation. Among the participants was a fellow Canadian, bridge designer Roger Dorton. With a major bridge project taking shape, Dorton was more than familiar with Davenport's work, particularly a seminal paper titled "Buffeting of a Suspension Bridge by Storm Winds." This paper picked up where Davenport's PhD thesis left off. Therein he had gone to the trouble of conducting a full-scale experiment of sorts at England's vintage Severn Railway Bridge, simultaneously measuring wind speed at several points along the lengthy 4,162-foot crossing, comparing and contrasting the varying speeds at the same point in time, and matching them with his own theory on the correlation of wind, as well as his turbulence spectrum, later to become known as the Davenport wind spectrum. (See sidebar, "Components of Davenport's Wind Theory," page 142.) With this raw data, the most valuable of scientific commodities, he proposed a new theoretical method for examining the response of linelike structures to turbulent winds (linelike structures being a suspension bridge or a transmission line, a horizontal structure where the wind gusts are much bigger than the bridge deck or the cable). His "buffeting theory," as it was called, predicted the bridges' vibrations in various modes and proved that their dynamic motion resulted not only from their inherent structural instability or weaknesses but also from the nature of the wind's turbulence specific to a geographic region—a simple conclusion, but groundbreaking all the same. He then applied and refined this buffeting theory by testing it on design plans for the Forth Road Bridge, just under way in Scotland.

About three years after his PhD, in 1963, when he debuted on the scene at that first international conference on wind engineering in Teddington, Davenport set eyes on the nearly completed structure. In addition to a cocktail party and a river cruise, conference social events included a bridge tour—the proceedings noted that the day after the meeting concluded, "a small number of participants journeyed north to inspect the new Forth Road Bridge then under construction, and the Forth railway bridge." It was an appropriately windy day. Several distinguished bridge engineers, including Germany's legendary Fritz Leonhardt, joined the excursion. Davenport served as the de facto guide, with keen verging on reckless enthusiasm. He became intrigued by the road bridge's deck, with three longitudinal air gaps between the roadway and cycleway/footpath designed to improve the aerodynamic stability, and then once in the zone he became deleteriously entranced by an equally enthralling feature of the nineteenth-century rail bridge, leading him to a close encounter with a train. Davenport couldn't help himself when the overtures of a bridge beckoned. And three years later again, in Lisbon, he listened intently as Dorton divulged his plans for a suspension bridge on the east coast of Canada, linking the Halifax Peninsula with Dartmouth, Nova Scotia. Davenport's first question to Dorton was, of course, "Are you going to do any wind tunnel testing?"

Figures 75A–B. In 1963, at the first international conference on wind engineering, held in Teddington, Davenport led an excursion to visit the Forth Road Bridge, then under construction in the east of Scotland, spanning the Firth of Forth (above). The entourage of engineers also visited the Forth Rail Bridge (below, in background), where Davenport had a close encounter with a train. Courtesy of the Boundary Layer Wind Tunnel Laboratory.

Components of Davenport's Wind Theory

Davenport's Wind Correlation

Correlation of the wind on a bridge is a measure of whether or not the wind is in sync along the bridge's span. Is the wind in phase, is it the same magnitude, is it the same direction? If wind is 100 percent correlated—that is, perfectly correlated—then the wind speed, the wind direction, and the wind phase are identical from one point to another along the structure. For a bridge, this would bring unfortunate consequences, since the excitation of the modes would be identical all along the span and would compound one another. Conversely, if the wind speed, direction, and phase are 0 percent correlated—that is, uncorrelated—then the wind would hardly excite the modes of the bridge at all, since the wind effect at one point would be cancelled out by the wind effect at another point. Davenport theorized that wind correlation is subject to exponential decay with distance—100 percent correlated at zero distance and 0 percent correlated at infinite distance, decaying exponentially in between at some decay coefficient. The decay coefficient is a factor of ground roughness and the nature of the structure. In the case of a bridge, the decay coefficient would entail correlation in the horizontal plane; in the case of a building, it would entail correlation in the vertical plane. Davenport's measurements on the Severn Railway Bridge confirmed this theory and enabled him to quantify the decay coefficient of over-water conditions for the longitudinal component of wind in the horizontal plane.

Davenport's Wind Spectrum

A spectrum, $S(f)$, is the variation of energy with frequency, f. This is a mathematical quantity and was in use in many fields before Davenport adapted it for wind. He made it a simplified function that could be mathematically integrated or summed along the frequency axis. This was necessary in the days before computers, since the buffeting theory calculations had to be performed by hand and with a slide-rule. Davenport performed a fit to his measured spectra from the Severn Bridge measurements and came up with the following:

$$S_u(f) = 4.0\kappa \bar{V}^2 \frac{x}{(1+x^2)^{4/3}}$$

Where $x = 4000\dfrac{f}{\bar{V}}$ and $\dfrac{f}{\bar{V}}$ is in cycles/ft

$\bar{V}$ = mean hourly wind speed at 33 ft (10 m) height

κ = surface drag coefficient, related to roughness of terrain

Davenport recognized that when the spectrum is plotted as $(fS(f))$ on a logarithmic scale, the area under the spectrum (that is, integrated over frequency or measured graphically using an instrument called a planimeter) is equal to the variance, or σ^2, which then can be used to determine the turbulence intensity, $I = \sigma/\bar{V}$. The maximum point in the spectrum also then gives a measure of the characteristic wavelength of turbulence, or the length scale (see figure 76A). The turbulence spectra of the vertical component, longitudinal component, and lateral component all differ in their distributions of energy with frequency or wavelength. The lateral and longitudinal components concern vertical structures such as buildings and towers, while the vertical and longitudinal components concern horizontal structures such as bridges and transmission lines.

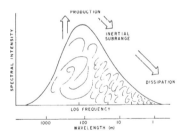

Figure 76A. Davenport's illustration showing the wind spectrum with three zones of atmospheric turbulence: the production zone, the inertial subrange, and the dissipation zone. The production zone, and its large eddies with large wavelengths, is the machine that generates the smaller-scale turbulence, transferring energy into little gusts. Courtesy of the Boundary Layer Wind Tunnel Laboratory.

Davenport's Buffeting Theory

The dynamic response of a structure as it resonates in one of its natural frequencies is called a "mode," that is, its vibration back and forth, in positive and negative directions, with a prescribed deflection called a mode shape. The dynamic motion of a structure when hit by wind (or by any dynamic load, such as an earthquake or highway traffic) is comprised of the combined motion in several modes of vibration. While a building, usually a basic cantilever structure stuck in the ground, has at most a few modes that are affected by the wind, the more complex constitution of a bridge can have eight or ten or more modes of vibration excited by the wind (all structures have hundreds of modes, but only a select few are important since they are excited by the wind). Each mode is excited by the energy in the wind at a different frequency. These are the bridge's resonant frequencies, the frequencies at which the bridge resonates with the wind.

A short bridge, for example, has higher resonant frequency. It needs more frequent gusts of turbulence to launch it into motion. But as the frequency of turbulence in the wind increases, it is composed of smaller and smaller and thus less powerful eddies, causing the energy in the wind to drop off (see figure 76A). As a result, these higher modes, and thus shorter bridges in general, are rarely excited at all. Conversely, a longer bridge with a lower resonant frequency is more likely to be excited, especially if its frequency happens to match the peak in the wind spectrum, the characteristic frequency of winds commonly seen in the geographic region in which it is situated.

As with his analysis of wind effects on buildings, Davenport recognized the importance of considering turbulence according to a bridge's mode shape, breaking down the turbulence into the various frequency components. Low-frequency background turbulence, for example, which affects a bridge in a static fashion, need not be of primary concern, whereas the turbulence associated with the structure's resonant frequencies is liable to excite the structure dramatically and needs to be more closely examined. In this sense, Davenport split up the wind loads, the buffeting loads, into background and resonant components. He separated out these components, analyzing each independently as warranted, and then combined them again at the end of the analysis to provide an overall unified picture.

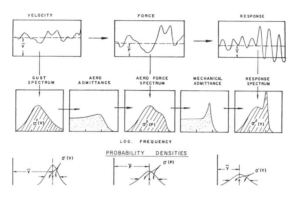

Figure 76B. Davenport's illustration showing the wind interaction functions—a variation on his wind loading chain. Courtesy of the Boundary Layer Wind Tunnel Laboratory.

In outlining this process in his buffeting theory, Davenport introduced various wind interaction functions (see figure 76B)—metrics for how the structure and the model interact with the wind, how the motion of the structure changes the wind, and how the wind changes the motion of the structure. A bridge, for example, moves up and down in the wind and rotates. As this occurs, the bridge becomes aerodynamically different from what it is when stationary. One of these wind interaction functions, the aerodynamic admittance function, had been used previously in aeronautical engineering but not in civil engineering. It was Davenport's development of the joint acceptance function and response or gust factor, in combination with the admittance function, that helped define how much of the wind's energy, at a certain frequency, is going to be picked up by the bridge and cause it to misbehave.

ധ   ധ   ധ

In the design stage, Roger Dorton hadn't planned on any wind testing. Nearing construction, he and the contractor, the Canadian Bridge Company, revisited the prospect, primarily because the A. Murray MacKay Bridge boasted a particularly innovative deck. The design called for a steel plate orthotropic deck—"orthotropic" being shorthand for "orthogonal anisotropic," meaning it possessed different stiffness in the longitudinal and transverse directions. The orthotropic double-duty deck, the first of its kind on a suspension bridge in North America, caused uncertainty about how the structure would behave in the wind, especially during construction.

Conventionally, a bridge has a lightweight concrete deck resting on a support system of trusses. Construction proceeds by first erecting the trusses without the deck in place—the deck flooring is an "add-on." This is advantageous because during erection, the wind blows through the lattice-like trusses; there are no big solid components to catch the wind like a sail and throw the bridge into instability before all the bolts are tightened and joints welded. Dorton's design, by contrast, with its orthotropic deck, meant that solid steel plates with welded stiffeners were an integral part of the truss system itself, added plate by plate as the truss went up. This system, too, had advantages. Since the orthotropic system provided extra stiffness, the depth of the stiffening truss could be reduced from 14 to 9.5 feet, in turn reducing the weight of the bridge (and resulting in an overall weight savings of 15 percent). But in the erection stage, the bridge would be at once lighter and a single solid prefabricated unit—all the better for the wind to grab it and blow it around. Pondering this aerodynamic quandary, Dorton decided it would be helpful to take an advance look at the bridge's behavior in the wind tunnel after all.

In light of these concerns about construction, the Murray MacKay demanded a more comprehensive wind program. Standard practice at the time was to test only part of a proposed bridge using what is called a section model—a representative portion from the center span, correlating with the real bridge and providing an abstract sampling of its behavior. But Dorton needed information both on the stability of the completed bridge and on its stability during various stages of construction. Davenport's tack, then, was to work not only with a section model but also with a full model of the bridge—the first full model of a bridge ever tested in turbulent boundary layer flow.

Tests unfolded in four phases, three stages of erection and one of the completed bridge, emulating the delicate choreography of construction: after the towers were in place, with their cables and hangers, the center of the main span was built, and from there the contractors worked simultaneously outward from the center and inward from the anchorages. Putting the full model as well as section models into play in the wind tunnel, Davenport discovered that the bridge at various stages of incompletion would be more susceptible to the wind; erecting the bridge with full decking would decrease the structure's critical wind velocity, resulting in instability at a lower velocity, and at a velocity below that of the completed bridge. The critical wind velocity varied according to each erection stage, with a minimum value of 47 miles per hour. Although not favorable, this did not warrant a new design, or even any alterations. When winds picked up, construction would simply have to come to a halt. This was valuable information for the contractors as it gave them confidence about what wind speeds allowed for safe working conditions. And it gave them a means to cost out the likely duration of construction, since knowing the wind speeds that would cause instability, Davenport

Figure 77. A time-sequence filmstrip of wind tests showing the A. Murray MacKay Bridge going unstable—revealing its "divergent response," its static instability in extreme winds. Courtesy of the Boundary Layer Wind Tunnel Laboratory.

was able to forecast, given the typical wind climate of the region, how many workdays the contractors could reasonably expect, and how often they might have to pack up and wait out the wind.

In evaluating the aerodynamics of partially erected suspension bridges, the Murray MacKay testing put the spotlight on an element of risk hitherto overlooked. Davenport also wanted to shed light on a number of research questions pertaining to the modeling process. He wanted to reexamine the conventional wisdom that section models were an accurate stand-in for full models. And, naturally, he wanted to compare the effects of smooth laminar flow versus turbulent boundary layer flow.

Dorton had taken a gamble in designing his bridge without wind tunnel testing. He carried out preliminary calculations and applied existing theoretical and empirical methods to estimate the critical wind velocity. He figured the completed bridge—with its specific design and structure, regardless of where it was located—would go unstable at winds of 130 miles per hour. Meanwhile, the regional wind velocity that Dorton designed for—the wind the bridge would likely face based on statistics of wind speeds for the Halifax area—was only 95 miles per hour (if he were designing the same bridge for Florida, it would be considerably higher). This left an ample safety margin of 1.38. Normally the goal is for a safety margin of 1.5—for the wind speed at which a bridge goes unstable to be 50 percent higher than that expected in the region over the lifetime of the bridge.

When Davenport began the wind tests, this looked like wishful thinking, or so it seemed initially. First he tested both section and full-aeroelastic models, subjecting them to standard smooth laminar flow. This yielded a critical velocity for the bridge of 103 miles per hour. So much for the safety margin. But then, as Dorton recalls, "Davenport saved the day." Testing with the full-aerolastic model, he introduced turbulent flow. This more realistic simulation of the wind pushed the bridge's critical velocity from 103 miles per hour to 210 miles per hour. Of course, Davenport had predicted all along that turbulent wind would increase the critical velocity, and Dorton was pleased to see that prediction come true. Similarly, just as uniform flow underestimated the speed at which the bridge would go aerodynamically unstable, Davenport discovered in testing the full model—

Figure 78. Davenport at the A. Murray MacKay Bridge with one of the contractor's staff (right), looking down from the top of the tower on the nearly completed bridge, and affording a view of the truss (in middle ground) that sits on the cable and holds the hoisting mechanism used for lifting deck sections. Courtesy of the Boundary Layer Wind Tunnel Laboratory.

and integrating the wind climate data—that the section model also acted as an underestimator. Or, to put it another way, the section model overestimated how much stiffening the structure required. The full model provided a more balanced and forgiving perspective. These findings were revelatory for Davenport and a relief for Dorton, who, with the wind report in hand, could confidently proceed without changing a single detail of his bridge's design.

With the A. Murray MacKay Bridge under construction, New York's Bronx-Whitestone Bridge was on its way to the Davenport lab for the first time. Early one morning in November 1968, a storm with gusts of 70 miles per hour raced through New York City. The Whitestone undulated vertically as much as 10 inches, forcing the bridge's closure during the morning rush hour. Some of the motorists stalled mid-bridge in the traffic jam abandoned their vehicles and ran for terra firma. While admitting that the Whitestone bridge tended to move more than others, the deputy chief engineer at the Triborough Bridge Authority tried to calm the panic in his comments to a *New York Times* reporter. "All suspension bridges sway in the wind," he reassured. "If they didn't have give, they would snap."

All the same, for creature comfort, this incident made clear that the Whitestone's wind excitation had yet to be adequately quelled— lesser winds had triggered reports of amplitudes reaching more than a foot and a half. And the Whitestone's creator, Othmar Ammann, was no longer around to reassure that "the britch is safe." He died in 1965, one year after completing his crowning achievement, New York's comparatively rock-solid Verrazano-Narrows Bridge, the largest and longest suspension bridge in the world for seventeen years, with its 4,259-foot main span and anchorages the size of cathedrals. Despite Ammann's departure, the Whitestone remained under the purview of his Manhattan firm, Ammann and Whitney. Senior partner and chief bridge engineer Herb Rothman took charge of diagnostics, reviewing the structural health of the Whitestone as well as that of many other bridges in New York and elsewhere, earning him the moniker "bridge doctor."

In fact, Rothman long had his eye on the Whitestone, walking its steel and climbing its cables, conducting spot checks for worrisome wear and tear, since the mid-1950s. In 1954, tests were conducted by F. Bert Farquharson at the University of Washington, also the site of tests on the Tacoma Narrows replacement bridge. As far as the Whitestone went, Farquharson advised it could be in grave danger in winds as low as 45 miles per hour. Every time a high wind came up, Rothman received a call from the Triborough Bridge Authority asking him to make an emergency checkup. Purportedly, during each visit he lost his glasses to the wind. Everyone's dissatisfaction grew. The old retrofits drew greater criticism as ever more sophisticated analyses produced increasingly worrisome results. Encouragingly, however, the vertical and torsional mode frequencies remained constant from storm to storm, regardless of wind speed or direction. And the vertical mode amplitude peaked in milder winds, paradoxically dropping as the wind increased. But the torsional mode amplitude never seemed to climax, continually escalating with the wind. "This is typical of a bridge made torsionally unstable by negative aerodynamic damping," Rothman noted at an American Society of Civil Engineers congress. "Significant damage in feasible very strong winds could not be ruled out," he said, adding in conclusion: "Still to be determined was the speed at which it would become unstable. Further study was needed."

For further study Rothman knew where to turn. In the late 1960s, his firm reviewed plans for the Sears Tower, and with Davenport conducting the wind study, they became fast friends. When not discussing the project at hand, Davenport bounced a few ideas off Rothman regarding a new method for testing bridges the lab had in the pipeline. Thus, when it came to pass that the Bronx-Whitestone Bridge needed help, Rothman told the Triborough Bridge Authority there were two options for a testing site: Jack Cermak's army-affiliated lab at Colorado State or his preference, Davenport's lab. Not wanting the U.S. federal government to get wind of what was going on, the TBA gladly endorsed Davenport.

Davenport's new ideas stemmed from his innovative Murray MacKay wind program, which motivated him to try for something better still, something more experimentally advanced, more nuanced, more precise.

A number of drawbacks dogged the traditional section models, as Davenport had shown. They were less than perfect in predicting full bridge behavior. The shortcomings owed in part to the fact that section models could manage only two-dimensional measurement. Just as the rigid stick models used for the World Trade Center registered only two sway modes, the bridge section models recorded only vertical and torsional modes, leaving the three-dimensional dynamic response out of the picture. And there was another encumbrance. Section models could not be tested in the more realistic turbulent flow because of the finicky technicalities of scaling. Full-aeroelastic building models tested in turbulent flow were at a scale of 1:200 or 1:400. Section models were at least a factor of 10 larger, at 1:40 or

1:80. Generating gusts of proportional size—gusts of several feet, say, versus the usual half inch or so—was nearly impossible with the current technology unless the lab built a wind tunnel about ten times bigger.

Full-aeroelastic model tests provided an alternative approach, though it too had drawbacks, the foremost being cost and complexity. "A full-aeroelastic model takes a lot of engineering to design the towers and the cables and the deck, to get everything modeled just so," says Peter King, a summer student at the lab during the seminal Murray MacKay testing, now a director. "And once you have the model built, you can't easily make any adjustments. While it might be possible to change the mass, you can't change the stiffness, and thus the frequencies, to reflect an evolving design." And oddly, the finely turned geometry of the towers and cables only adds entirely extraneous factors, obscuring the aerodynamic behavior of the key structural component—the suspended deck—and thus only serves to confuse comparisons with the more minimalist section model. But even using a full-aeroelastic model unto itself was not problem-free. In replicating the towers, cables, and deck, experimenters had to factor into their motion equations the key force working on the structural members: gravity. This required scaling with the Froude number. William Froude, a nineteenth-century English naval architect and engineer, formulated his namesake number by comparing inertial and gravitational forces to describe the resistance of water waves against ships in predicting their hull speed and stability. Fundamentally, the Froude number is a number accounting for the influence of gravity on motion, the ratio of the inertial to gravitational forces, and so the greater the velocity, the higher the Froude number. In the case of a bridge, it represents the influence of gravity on the load-bearing cables and their effective stiffness due to the tension in the cable when deformed in the wind. Factoring in the Froude number has the effect of restricting testing to awkwardly low wind speeds (the wind speed scaling is then equal to the square root of the length scale), which means the model won't move very much, and the electrical signals from instrumentation fixed to the model will be small, and on the whole it will be extremely difficult to measure the model's response. Low wind speeds also make more complicated the scaling of the air itself. Ideally, since models are a scaled-down representation of real life, the air should be scaled down as well; it should be less viscous. But in a typical atmospheric pressure boundary layer wind tunnel, it is impossible to generate the properties of the air exactly to scale (some aeronautical wind tunnels have solved this problem by pressurizing the air in the test section, or by using a different fluid such as cold nitrogen). The discrepancy is characterized by a parameter that defines the ratio of the inertial to viscous forces in the flow, the Reynolds number, named after Osborne Reynolds, another nineteenth-century English innovator in the field of fluid dynamics. When the Reynolds number is large, the inertial effects dominate, and when it is small, the viscous forces are the larger ones. Model tests in the boundary layer wind tunnel are performed at Reynolds numbers several orders of magnitude smaller than in full scale, but for most sharp-edged bodies the effect of this mismatch is small.

To wit, complexity, and thus expense. Adding insult to injury, it was superfluous complexity—there was no value added. There had to be a better way, a way to fabricate a more minimalistic model while still maximizing the quality of the information produced. Davenport wanted to marry the best of both the section model and the full model. He wanted a three-dimensional model that could be tested in turbulent flow, though minus the fiddliness of the Froude number, and a model that would be more abstract in its representation of the three-dimensional dynamic response of a bridge.

Davenport reasoned that the part of the bridge most sensitive and responsive to the wind is not the cables or the towers. Rather, it is the suspended component, the deck—long and wide, like the wing of a Boeing 747. But he preferred to envision the deck as a simple linelike structure, like a clothesline. When free of laundry, the line is light, vulnerable, susceptible. When hung with heavy wet clothes, the mass makes it taut and stable. With this linelike structure in his mind's eye, Davenport's breakthrough came when he realized that a bridge's mode shape, its curvature or displacement caused by vibration, was sinusoidal, exhibiting the standard sine wave, the undulating line on the graph so common in high school math classes. This was the mode shape of any basic line structure, whether a clothesline, a piece of string, a skipping rope, or a strip of ribbon. And this became the basic building block for his new testing method, which he named aptly the "taut strip"—reducing a bridge to its core, its essential component, the main strip of deck, eliminating the extraneous towers and cables and side spans. So long as Davenport could take into account the effect of the cables and towers in another way, he could dispose of the gravitational factor; he could remove the finicky aspects of scaling from the equation.

Elegantly spartan, the taut strip model comprised two taut wires strung between anchor blocks. Davenport found that piano wire was the strongest wire of the requisite grade and size. Stainless steel tubing also worked well and provided additional torsional stiffness. In terms of what it sought to represent, the taut strip model was halfway between a section model, modeling only a slice of the main span, and a full model, modeling the entire bridge. The taut strip model replicated the entire main span, with mass- and inertia-scaled deck cladding segments fixed to the top of the wires. Varying the spacing and tension of the wires, Davenport fine-tuned the torsional, vertical, and horizontal stiffness to match that of the bridge in question. "Change the tension of the wires, and you change the natural frequency," says King. "As you increase the tension, just like a guitar string or a piano string, the natural frequency goes up." And the farther apart the wires, the stiffer the system in torsion. With this minimalistic makeup, Davenport created a model that could mimic the mass and inertia and wind-induced movement of the bridge's main span. "He pared the bridge down to its most essential building blocks," says King. "That's what Alan did so well in everything he touched, really paring down to the essentials. If the tower doesn't have anything much to do with the dynamics of the bridge, then why should you model it?" The taut strip model achieved Davenport's goal: an economical hybrid between the section model and the full-aeroelastic model. And it bore financial benefits. The bill for a taut strip model ran to less than half that of a full-bridge aeroelastic model. These financial savings translated directly into time savings. Built more quickly and tested more quickly, the taut strip got results faster for clients.

Early on, Davenport deployed the taut strip model informally during pure research investigations, driven by professional curiosity. He examined the Golden Gate Bridge, proving that the taut strip model was capable of confidently simulating full-scale motion (completed three years prior to the Tacoma Narrows Bridge, the Golden Gate Bridge was another Moisseiff bridge that displayed wind sensitivity). And he reconstructed the original Tacoma Narrows Bridge. It made a particularly nice test case since the structure went unstable at such a low wind speed and since it was perhaps the most tested bridge on the planet, providing lots of data with which to compare results. For these explorations he worked for no client, he searched solely for engineering edification. The lab, in this respect, was the bustling center of a Venn diagram—or what Davenport fondly referred to as the lab's three-ring circus, with intersecting spheres of research, education, and commercial application (though Davenport's curriculum vitae listed a deliberate "Nil" under "Patents"). As for commercial applications, when New York's bridge doctor appeared on the scene demanding the latest in aerodynamic bridge testing technology, the taut strip found its first taker. Rothman had a problem bridge on his hands with the Whitestone, and the conservative estimates generated by conventional wisdom would not help his case.

Figure 79A. In 2000 the lab constructed a full-aeroelastic model for the Bronx-Whitestone Bridge. Courtesy of the Boundary Layer Wind Tunnel Laboratory.

Figure 79B (top). Laser deflection instrumentation under the Bronx-Whitestone model's bridge deck. Courtesy of the Boundary Layer Wind Tunnel Laboratory.
Figure 79C (middle). Accelerometers located at the top of the Bronx-Whitestone model's towers. Courtesy of the Boundary Layer Wind Tunnel Laboratory.
Figure 79D (bottom). Pots of water with plungers increased the damping of the Bronx-Whitestone model, mimicking the effect of the bridge's tuned mass damper. Courtesy of the Boundary Layer Wind Tunnel Laboratory.

ᴝ ᴝ ᴝ

Beginning in the early 1970s, the Whitestone went through an intensive battery of tests at the lab. Davenport made several visits to the bridge, familiarizing himself with the structure's scale and form, questing for new information that could be factored into the wind program, and, if the winds were blowing in his favor, gleaning some firsthand experience of how the Whitestone moved. On one occasion he brought along a posse of engineering students and arranged for them to visit another Rothman charge, the more windworthy Verrazano-Narrows Bridge (Rothman thoughtfully closed the bridge to traffic temporarily, allowing the students to inspect the deck with police escort). And simultaneously with the various visits, Davenport subjected the bridge to the full arsenal of wind testing tools: a taut strip model, as well as a section model, and later a full-aeroelastic model.

The program proceeded as follows. First, for each major structural element, Davenport determined the vibration amplitude that would cause both the onset of damage and major damage. Surprisingly enough, tests showed that the cables and suspenders could endure over 20-foot amplitudes at the center of the main span, while the towers and stiffening girders could survive amplitudes between 12 and 16 feet. The most vulnerable component was the lateral system. Imagine the deck, made of two side girders and a concrete slab sitting on top, the lateral system being the X-bracing between the two girders. In this case, the X-bracing was right in the middle of the girders, with nothing beneath. This formed a boxlike structure with no bottom, making it floppy and weak and susceptible to damage at amplitudes of less than two feet. Next, Davenport determined what winds would cause these damaging amplitudes. And then, using meteorological records, he calculated the chances of those damaging winds occurring in the region, in terms of both wind speed and wind direction. This last factor, wind directionality, was a crucial factor in concluding that the Whitestone was not as vulnerable as expected, with the return period for the critical wind speed of 60 miles per hour assessed at a respectable fifty years (whereas without wind directionality factored in, the only conclusion possible was that the bridge had to be torn down before it fell down). At long last, the windy ways of the Whitestone became aerodynamically known commodities. "It had been such a mystery, up until then," says Rothman. Yet solving the mystery was at once reassuring and unsettling. Because fifty years, while a respectable length of time, was not nearly long enough. As Rothman succinctly summarized, there was still "a slight probability" that the bridge could collapse in a windstorm.

With yet another retrofit necessary—following the suspender-like stays and sightline-obscuring Warren trusses—the question was, what to prescribe next, what course of treatment remained? The cables were such a dominant structural component that stiffening any other aspect of the structure seemed futile. Changing the cable itself, Rothman felt, was impractical. And altering the overall weight of the bridge he deemed hopeless. Only two options remained: streamlining the bridge or increasing its damping. Davenport was in favor of the latter, and Rothman agreed. A damper would be far less expensive, its installation would not interfere with traffic, and the Whitestone's appearance would be preserved.

Rothman immediately started brainstorming possible damper designs. His first idea harnessed the power of the wind in a rather unexpected way. "The wind puts the energy into the bridge that makes it move. We needed to take the energy out of the bridge and put it back into the wind," he recalls. "What I was thinking of doing was installing a propeller that the wind would drive. The propeller would drive an electrical device, and the electrical device would drive electricity into heaters underneath the bridge. A bunch of wires underneath the bridge would be toasters, in effect—the wind would blow across the toasters, and toast itself." This quixotic notion of a propeller-toaster damper did not get very far. Going back to the drawing board, Rothman set out the damper's most basic and crucial criteria. It had to be reliable, trustworthy, unfailingly dependable. The device could not use any power source except wind—electrical failures, empty fuel tanks, and dead batteries were all Murphy's law contingencies that would have dire consequences during a storm. The damper had to start, run, stop, and restart all without any intervention. Consulting Davenport, Rothman opted for a tuned mass damper. "At the time, the one tuned mass damper that I knew of was for the Citicorp Building in New York," he says. He added that there was an urban legend that Citicorp occasionally showed off its computer-controlled damper by activating the device when there was no wind. "They would drive it and make the building vibrate, just for fun." The Whitestone damper would be entirely self-driven, excited in concert with the bridge when the bridge was excited by the wind.

Since it was intended to counter torsional twisting of the deck, this damper was more complex than most. Four 214-kilonewton weights (each about 20 tons, or the weight of a large sperm whale) staked out a square, a pair on either side of the bridge, 39 feet apart at midspan. Set into action whenever the deck motion exceeded a certain threshold, the weights began a dialogue of countermotion—motion that was tuned to be precisely out of sync with the bridge's motion and relayed through a network of cables, pulleys, and a 200-foot long torsion bar. "You had this thing moving," recalls Rothman, "—that was the energy input from the wind, kinetic energy, and now the question was how do you get the energy out of the damper?" The best bet for an energy-dissipating device turned out to be truck brakes, also known as disc brakes. Attached to the pulleys, they dissipated more than 200 horsepower. And with that, the tuned mass damper increased the return period for damaging winds from 50 years to more than 10,000.

ω   ω   ω

Tending to the antique and wind-weary Whitestone occupied a good amount of the lab's attention through the 1980s, but this rehab work on a relic was balanced out by demands of not one but two high-tech designs—one in concrete, one in steel—for Florida's new Sunshine Skyway Bridge.

The original Sunshine Bridge, a cantilever bridge built in 1954 crossing Tampa Bay, suffered irreparable structural damage when the inbound freighter *Summit Venture* collided with a support pier during a rainstorm in May 1980. Nearly 1,000 feet of the southbound bridge fell into the bay, killing thirty-five people. Governor Bob Graham vowed to make the replacement an iconic structure, a "flag bridge" for the state of Florida. This bestowed upon the state's department of transportation a mandate for the best of the best design.

"There was a very strong emphasis on the new bridge being a pioneer for new standards in the country, and the world," says Linda Figg. Her father, Eugene Figg, together with Jean Muller (of Figg & Muller, now Figg Engineering Group) put forth the winning design for a prestressed concrete cable-stayed bridge, prevailing over the steel alternate. Both the concrete and the steel options were cable-stayed designs, and both were tested at the Davenport lab. Given Graham's imperative for excellence, each design came with a sizable pot of money—the means to ensure that the longer, higher replacement bridge would be strong enough to withstand freighter ship impact, not to mention being aesthetically unrivaled and aerodynamically sound (Florida and the Gulf Coast are the bull's-eye for Atlantic hurricanes). Davenport had experimental carte blanche in executing a state-of-the-art wind program for a state-of-the-art bridge.

By that point in the twentieth century, the evolution of the science of wind engineering mostly amounted to advances in calculating wind loads on tall buildings—Davenport's gust factor for computing wind forces on buildings had been integrated into structural codes. Bridges, meanwhile, lagged behind. Bridge designers deferred to the traditional code of practice, burdening their bridge with more steel than necessary to make sure the structure stood strong. A wind tunnel lab was not yet considered a design tool; bridge clients hired Davenport as a consultant after the fact. With the Skyway, designers willingly ventured into uncharted waters.

Davenport proceeded by developing a series of special tests that would allow him to tease out more empirical data from the wind tunnel, which in turn would provide a more nuanced assessment of wind loads on the full-scale structure.

He focused his efforts on the section model method. Prior to that, the information gleaned from the two-dimensional static section model afforded a basic appraisal of the static forces acting on the bridge section and whether or not a bridge was stable. This information in turn was massaged analytically with Davenport's theoretical wind structure interaction functions to determine whether the loads were in the ballpark of design specs. This was a laboriously complicated process, and woefully unsatisfactory—another process Davenport was determined to improve upon.

Figure 80. Representatives from the American firm Greiner Associates and the German firm Leonhardt und Partner closely inspecting the section model of the steel alternate for Florida's Sunshine Skyway Bridge. Courtesy of the Boundary Layer Wind Tunnel Laboratory.

Together with King, by then a research associate, Davenport figured it should be possible to get more information, greater detail, from a section model, information such as the wind interaction functions—the building blocks in his wind loading chain, the effect of the mode shapes, the natural frequency, the correlation, the aerodynamic admittance function—that heretofore had only come into play theoretically. "Up until then, the only component measured in the wind tunnel was the force, the force the model experienced due to a certain wind speed. This was first time we actually tried to measure those other aerodynamic functions in the wind tunnel," says King. "The goal of the teasing out was to verify the theory through measurement in the wind tunnel, to measure what these individual functions were, rather than just the total gross effect." If wind metrics such as the wind interaction functions could be defined in practice rather than just in theory, wind loads could be measured in more detail.

More specifically, the additional detail Davenport was after would help him define a bridge's response—that is, the mode shapes the structure took when excited and deformed by the wind. While a building, a simple cantilever structure stuck into the ground, usually has at most three mode shapes excited by wind, a bridge, with its more flexible structure, has many mode shapes, and all of them are important in relation to the wind. "Take a steel ruler," says King by way of explaining, "clamp it to the desk, pluck the end, and it's going to move only up and down, that is the first mode of vibration. For the second mode, you'd have to hit the ruler really hard, and then it would move like a question mark, with a double curvature. In the case of a bridge, it's more like a guitar string with different harmonics. So you've got the first mode, which would be a half sine wave— maximum in the middle and zero at either end. And then the next harmonic is a full sine wave, a node point or a zero point in the

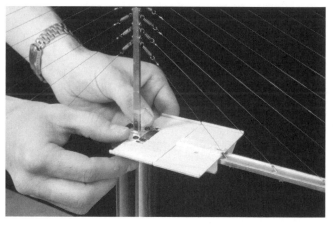

Figure 81. The internal structure of the full-aeroelastic model for the Sunshine Skyway's concrete alternate, with cladding sections mounted piece by piece on the deck spine. Courtesy of the Boundary Layer Wind Tunnel Laboratory.

Figure 82. The lab helped with the design of dampers for the Sunshine Skyway's cables. The triangular struts run off the cables and down beneath the bridge, and into shock absorbers that reduce motion. Courtesy of the Boundary Layer Wind Tunnel Laboratory.

middle. And then the third mode, fourth mode, fifth mode, would break up the sine wave into higher orders, 1.5 sine waves, 2 sine waves, 2.5, and so on. Those are the mode shapes." Again, as simple as it seems, this was another of Davenport's insights: that wind loads would vary along the length of the bridge according to mode shapes. For instance, the major load a bridge carries is the dynamic load incurred during torsional motion, and this load in turn varies according to a bridge's mode shapes. It only made sense, then, that loads should not be applied to the structure as one homogenous unit. They should be ratioed, or rationed out, according to the mode shapes,

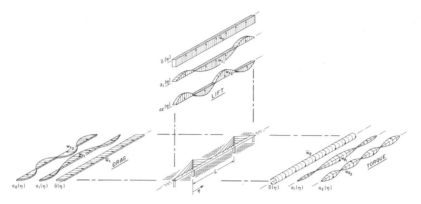

Figure 83. The three types of modal loads—drag loads, lift loads, and torque loads (shown from left to right)—all act on a bridge deck to different extents at different points along the structure. For the fundamental modal load, larger loads occur at midspan than over the piers, since a bridge moves more at midspan than it does at the piers. As a result, the load is not uniform from one end of the bridge to the other. In the figure, the uniform load from one end to the other is not a modal load but is the mean load, which is always present. The remaining six sinusoidal loads (the symmetric and antisymmetric shapes for each of drag, lift, and torque) are static equivalent representations of the dynamic loads and are shown at their maximum amplitudes. These are the modal loads, which are applied to the structure one at a time. The effects of the loads (stresses, deflections, etc.) are combined as a square root sum-of-squares because the modal loads do not occur simultaneously but rather are random. Courtesy of the Boundary Layer Wind Tunnel Laboratory.

according to the movement of the bridge. "Take Newton's second law, that force is mass times acceleration," says King. "If the bridge is not moving, or not moving at a certain spot, meaning it is a node point, then there is no force or load. But if it is moving, and you know what the acceleration is, and you know what the mass is, then you know what the force is, and that's the equivalent static load created by the wind." Breaking it down into the different accelerations in the different modes gives modal loads that are proportional to the modal deflections. And naturally, there are larger loads at midspan than over the piers since the bridge is moving much more at midspan than at the piers. "It's not a uniform load all the way from one end to the other," says King. "It varies."

By teasing out more empirical data from the wind tunnel, Davenport could capture this variegated modal behavior of the bridge. To do so, he engineered a few amendments to the traditional procedure for testing a section model in turbulent flow. For one, he reduced the problems associated with the scaling of the wind correlation by extending the length of the model, focusing on a longer portion of the bridge. He also advocated for a larger geometric scale in order to model the important aerodynamic features of the bridge deck. And he managed to get a direct readout of the local vertical forces and rotational torsional forces on a narrow "slice" through the bridge deck by installing pneumatically averaged pressure taps around the perimeter of the model, another technique recently invented at the lab (see the sidebar, "Averaging the Air," page 204).

With these tweaks, Davenport got the extra information he was after. He quantified and confirmed the previously purely theoretical wind interaction functions, metrics that in turn more accurately quantified the wind loading on the full-scale structure. Such intricate and accurate measurements hadn't previously been possible. Standard practice had been to measure the overall effect without defining the building blocks or individual components. Now Davenport was able to take the measured behavior of the section model, deconstruct the various building blocks of that behavior, and develop a pared-down formulation for loads—equivalent static loads, as he called them, a shorthand of sorts for turbulent wind loads, expressed in terms of static wind loads (since nobody knows how to apply a dynamic load in design). "Most bridge engineers want one load that they can put on a bridge that would cover everything about the wind," says King. "But you can't really do that. For every bridge there are several equivalent static loads. What we've done is broken the overall wind load down into the various modal wind loads—mode 1, mode 2, mode 3—and this is what the engineer gets in a wind tunnel report, together with a statistical synopsis accounting for the combination of these loads, since they're not likely to reach their maximum value, all at the same time." Engineers were accustomed to dealing with combination loads in this way, because that's how they dealt with traffic loads, and earthquake loads, and usually they had just one wind load. As specified by codes, engineers would combine these loads based on their likelihood of occurrence—there might be 100 percent of the traffic load and only 50 percent of the wind load, since it would be unlikely to have the design traffic load and a hurricane on the bridge at the same time. "But engineers weren't accustomed to breaking wind load down in this fashion—50 percent of mode 1, 30 percent of mode 2, and 20 percent for mode 3—if, for example, they were designing the bridge deck at midspan, whereas there would be a different combination if they were designing a specific cable or component of the tower," says King. "They understood the process. It was just a bit more work. But in return it gives them a more accurate picture of what the effect of the wind loads will be on their structure."

And the bottom line, again, with these quintessential Davenport simplifications was that bridge testing became more realistic and more reliable. Just as with a tall tower, it wasn't sufficient, desirable, or even advisable to take a uniform wind force associated with the largest anticipated wind speed and apply that during design. The goal was to determine actual loads that the bridge was likely to experience. In reaching that end, the equivalent static load method was a complementary approach, further enhancing the wind engineer's repertoire of tests and techniques—the range now ran from static section model tests in smooth flow and turbulent flow, to these new dynamic section model tests in smooth and turbulent flow, to the taut strip model, and finally the full-aeroelastic model. Ammann and Steinman surely would have welcomed such progress in the bridge-builder's art, deepening the engineer's probative powers and yielding results ever more profound and precise. No longer was the wind tunnel relegated to the back end of design. Now it was a tool that offered valuable insight from the outset.

With the Sunshine Skyway Bridge, this enhanced process ultimately influenced the structural system selected—a large box girder with streamlined elliptical piers—and resulted in a bridge impervious to severe winds. Opened in 1987 and officially named the Bob Graham Sunshine Skyway Bridge, the 5.5-mile structure has a cable-stayed main span that runs to 1,200 feet—the longest concrete cable-stayed bridge in the world when it was built—with an arresting, radiating spray of cables painted a radiant solar yellow. And though it is often praised for its aesthetics, the real proof is in its structural per-

Figure 84. The Bob Graham Sunshine Skyway Bridge, which in part was inspired in its design by a visit then-Governor Graham took to France, home to many a cable-stayed bridge. Delmas Lehman/Shutterstock.com.

formance. "We can see the proof," says Linda Figg. "The bridge has gone through many, many hurricanes since it opened, some of the largest hurricanes to hit Florida"—among them many of the boldface hurricanes of the last two decades: Andrew, Gordon, Allison, Erin, Opal, George, Floyd, Irene, Frances, Jeanne, Wilma, and Katrina. "The Skyway has always stood strong, there has never been any damage or concern," says Figg, adding, tongue in cheek: "I always tell people, 'If there is a hurricane coming and you want to be in a safe place, go inside the Skyway bridge'"—inside the hollow concrete girder box that forms the deck, where only maintenance staff have access—"because I know, given Dr. Davenport's wind science technology, that it is secure."

In September 1999, President Clinton declared a state of emergency, and more than one million Floridians evacuated their homes as a

Category 4 hurricane charged toward landfall. Making an abrupt turn and traveling northward parallel to the coast, Floyd approached New York City a day later, closing public schools with the promise of torrential rains and high winds. The Bronx-Whitestone Bridge was more than ready—ready to measure the storm, at the very least.

One year earlier, inspection of the Whitestone's main cables, tipped off by acoustic sensors registering the sounds of wire breakage, revealed unacceptable levels of deterioration, compromising safety standards. The bridge was showing its age. Hefty upgrades over the years had taken their toll, what with the stays, the trusses, the mammoth 100-ton tuned mass damper, and the increased loads from more and more traffic. It was now the structure's overall weight, not so much the wind, that threatened the bridge's life span. Rothman, by then at the firm Weidlinger Associates and nearing retirement, again turned to the Davenport lab. With advancing technology, the restoration options had moved beyond the brute force methods of the past. Fifteen years earlier, Rothman dismissed streamlining the bridge aerodynamically or reducing its weight as impractical and hopeless. Now those avenues were feasible. While the Weidlinger team reevaluated the situation, the lab suggested that a savvy and prudent first step in the treatment plan might be to put the bridge under full-time, full-scale monitoring to get the best possible measure of its vital signs.

And so as Hurricane Floyd blew into New York on a mid-September morning, the marine anemometer spun into action, perched 10 feet above the bridge on one of the lamp standards. Three accelerometers registered the bridge's vertical and lateral accelerations. Six strain gauges monitored the trusses, quantifying their role in the bridge's stiffness or exposing their impotence. And secured to the bridge's maintenance catwalk resided the brains of the operation, an industrial-strength computer capable of withstanding Arctic temperatures. The computer's data logger translated the wind and bridge action into binary bits and bytes. It was a "smart system," meaning the software logic was coded to continuously scan all channels of data for any action. Before archiving in the data registry, the logger transformed the raw information into a spectrum of statistics, computing the maxima, minima, mean, and root mean square values for ten-minute samples. And when three consecutive 10-minute periods surpassed the threshold of interest—25 miles per hour, or, by the Beaufort scale, a strong breeze that bends tree branches into motion, whistles telephone wires, and renders umbrellas inside out—the data logger switched to time-history mode, recording all channels constantly for a 90-minute period. The system's 40-MB memory could retain eleven such 90-minute sessions, or 16.5 hours. Hurricane Floyd triggered 12 hours of continuous wind history.

The computer transmitted the data two miles north to the bridge office and into the Bridge Authority's network. From there the logger underwent nightly Internet interrogations by a computer at the lab that downloaded the data for analysis. By the time the hurricane hit New York City, it had weakened to a Category 2, capable of 109-mile-an-hour winds. But Floyd peaked at 60 miles per hour, with mean wind speeds reaching 46 miles per hour, driving the bridge's torsional accelerations to 20 milli-g's at the girder edge for 10 minutes, enough movement to induce some nail biting among nervous bridgegoers but not overly alarming. It wasn't the severest of storms. Nevertheless, in recording this high-definition storm profile and measuring the impact on the bridge, the lab had once again managed to capture a rare opportunity for full-scale monitoring.

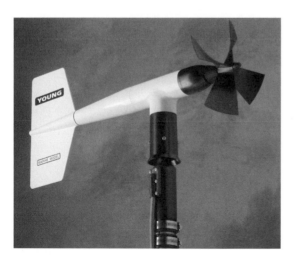

Figure 85A. For the Bronx-Whitestone Bridge's continuing regime of full-scale monitoring, two anemometers (R. M. Young Marine Anemometer Model 5106) are cantilevered from the deck, recording easterly and westerly wind speeds. Courtesy of R. M. Young Company.

Figure 85B. One of seven accelerometers installed on the Bronx-Whitestone Bridge, each measuring vertical and lateral accelerations. Subtracting a vertical acceleration on one side of the bridge from a vertical acceleration on the other side gives the all-important torsional acceleration, which monitors for any Tacoma-like instability. Courtesy of the Boundary Layer Wind Tunnel Laboratory.

Figure 85C. During Hurricane Floyd, the data logger, secured to the bridge's maintenance catwalk, transmitted wind data two miles north to the bridge office, into the Bridge Authority's network, and from there it underwent nightly Internet interrogations by a computer at Davenport's lab. Courtesy of the Boundary Layer Wind Tunnel Laboratory.

The accrual of full-scale data, as Davenport described in 1999 at the 10th International Congress of Wind Engineers in Copenhagen, was becoming an increasingly invaluable tool in helping engineers wrangle the outstanding issues in bridge aerodynamics. Though the purpose of full-scale monitoring, he cautioned, was not so much to measure whether the theoretically or experimentally predicted behavior of a bridge held true in reality, especially since one might wait tens or hundreds of years for the design wind to occur. Neither was full-scale monitoring tantamount to testing in a gigantic wind tunnel. Instead, he felt this line of inquiry should be used in ironing out the remaining wrinkles in theoretical models. Theory is innately embedded with uncertainties and unknowns, blurry factors that scientists perpetually strive to pull into sharper focus. Full-scale monitoring was a crucial means of further pulling apart and refining the building blocks of theory—each building block containing its own uncertainties, some larger than others. For example, material properties, such as stiffness and strength, are reasonably well-known commodities, since there has been a lot of research on the strength of materials and since materials' performance also depends on quality control during construction. The CN Tower and Confederation Bridge (linking Prince Edward Island and New Brunswick), both vetted at the lab, were structures with very high-quality materials, and thus with very little uncertainty regarding their structural properties. Perhaps not surprisingly, the nature of wind itself is still the building block of greatest variability.

For the Whitestone wind tests, the lab gathered long-term records from LaGuardia Airport and translated this data as best as possible into site-specific information using theoretical models. But here, still, basic questions dogged Davenport: *How accurate were those theoretical models in actually replicating the wind? How does a stationary condition in the wind tunnel compare to real life? How long must a storm be present and stable for the amplitudes of motion of a bridge to reach a constant level? How long does it take for these to reduce once the wind speed or direction changes?* A few years of monitoring the wind at the actual site would quantify these uncertainties very easily. "Then one doesn't have to wait for 100 years for the answer to be evident," explains King. "In a wind tunnel, where everything is very controlled, you can force something to go unstable, but how long would that instability take to occur in real life? It won't happen instantly, as soon as the wind speed reaches a certain value. It takes a while to build up over time. It may take 15 or 20 minutes, an hour even. By doing the full-scale monitoring you can see how the bridge really responds to varying wind speeds and directions, how the response increases with time, and then you can extrapolate to determine how long it would actually take for instability to occur." The full-scale line of inquiry, without a doubt the most realistic and reliable tool of all, augments the collective repository of data and guarantees that the theories and research methods are advancing.

With the Whitestone under constant observation, next came the retrofits. First and foremost the bridge needed to lose weight. "We put this bridge on a diet," the president of the Triborough Bridge Authority told the *New York Times*. "Here's a bridge that's 65 years old. It got a little heavy around the midsection. Just like with human anatomy, as you get on in years. In this case, instead of putting an extra strain on your heart and other organs, it's putting a strain on the supporting structure, the skeleton of the bridge."

The lab ran dozens of options through the wind tunnel, profiling the rehabilitated bridge should it undergo retrofit A, B, or C, trying to determine what the best aerodynamic fix would be, and at what cost. One scenario entailed removing the stiffening trusses, which raised questions as to how the bridge would hold up on its own. "The answer was not good," says Genaro Velez, who took over as Weidlinger's Whitestone project manager (Rothman having retired, but ever keeping a close eye on the bridge's progress). The trusses could be removed, but not without souping up the bridge aerodynamically to compensate, and on that score, a prescription for wind fairings won out. Just as a race car's spoiler system of fins is meant to "spoil" unfavorable airflow across the vehicle, these aerodynamic devices on bridges reduce drag and lift. In aeronautics, such a device is often called a "lift dumper," which is exactly what was necessary on the Whitestone. The lab tested a number of configurations for fairings: curved like a half round or triangular; made from steel or forged from fiberglass. All the fairings tested had worked well with other structures. The trick was finding the fairing that worked best with the Whitestone, deflecting the wind over the structure in a steady stream rather than exciting it with subsidiary eddies, with by-product buffeting, or even with fatal flutter, causing instead of curing in-

Figures 86A–B. For the Bronx-Whitestone Bridge's fairings, a triangular form (left) was found to be most effective in "spoiling" the unfavorable lift. The installation of the fairings went section by section: once a section of fairing was installed, the truss above could be removed (right). Courtesy of Weidlinger Associates | MTA Bridges and Tunnels.

stability. After the lab determined a triangular contraption was the best fit, tests continued on small triangles and large triangles, triangles with a flattened top corner and a flattened bottom corner. Finally, the lab recommended a fairing in the shape of an isosceles triangle: a wedge with two sides nine feet long and projecting seven feet off the structure's girders for the entire run of the bridge. At a cost of $32 million, Weidlinger removed the trusses and installed the fairings in stages. But not without continued full-scale monitoring, the engineers keeping close surveillance on how the Whitestone and the wind settled into their new rapport.

With the fairings in place by April 2004, the Whitestone's retrofit entered phase two. The lab had studied redecking the bridge with an orthotropic steel deck much like that of the A. Murray MacKay Bridge. The original concrete grid deck sat on the subfloor beams and served only as a roadway, contributing nothing to the stiffness of the bridge, nothing to the resistance of torsional vibrations. The lab

Figure 87. With the installation of the new deck in 2008, the Whitestone's transformation was complete. Courtesy of Weidlinger Associates | MTA Bridges and Tunnels.

Figure 88. The fully restored Bronx-Whitestone Bridge. Courtesy of Weidlinger Associates | MTA Bridges and Tunnels.

tested decks of varying stiffness and shapes and assessed lowering the lateral system from the midpoint of the girders to the lowermost flange, incorporating the deck as the top member of the structural system. This new deck would work with the lateral system to stabilize the span against wind by forming a "torque tube box." Before, the bottomless bridge was an open box, floppy and twistable. Now, forming a closed box, it was much more resistant to twist.

With the installation of the new deck in 2008, at a price tag of $136.7 million, the Whitestone's transformation was complete. The bridge lost 6,000 tons or 25 percent of its dead weight. Its critical wind speed increased from 60 miles per hour to 138 miles per hour. Yet the full-scale observation of the bridge continues—and with an enhanced instrumentation system that includes dual anemometers and several more accelerometers. The lab submits reports to the Triborough Bridge Authority every six months, though there is nothing much to report until the next Floyd hits, providing real-time, real-life feedback on the effectiveness of the retrofit. "What we need is a really good storm," says Velez. With the life span of the bridge lengthened by decades, if not centuries, he is confident the Bronx-Whitestone Bridge is now fit as a fiddle to withstand the worst wind forces for the foreseeable future. "I'd tell you it should last forever . . ." he says, but then he stops short, hedging his bets.

Aside from the historic Whitestone, Davenport's professional pilgrimage from bridge to bridge over the years filled his curriculum vitae with the most futuristic of structures. He was tickled to serve with the eminent German structural engineer Jorg Schlaich on the French government's commission for the Millau Viaduct, seeking confirmation that the wind program conducted in France was up to snuff (and flooding Davenport's basement with boxes upon boxes of Millau reports). Warning of wind vulnerabilities during construction, Davenport otherwise gave the go-ahead for France's largest overland bridge—then the highest road deck in the world, at 886 feet above the Tarn River, and boasting the tallest pylon, at 1,125 feet, taller than the Eiffel Tower. Likewise, France's Pont de Normandie benefited from the lab's arm's-length expertise, with King sent as an emissary to the Centre Scientifique et Technique du Bâtiment, parlaying his experience with taut strip tests. Denmark's Storebælt Bridge, too, utilized the taut strip's expedient experimental method. Its wind program took on feverish intensity at the lab as the structure vied for the title of the world's longest suspension bridge. When it opened in June 1998, the 5,328-foot main span missed the mark by mere months. Japan's Akashi-Kaikyō Bridge, completed two months earlier, with a span of 6,532 feet, clinched the record for the longest suspension bridge arching over the planet.

Davenport had visited the Akashi-Kaikyō in 1998 with Sheila while attending a symposium on long-span bridges and high-rise structures in Kobe. There transpired some interesting if collegially competitive discussions between Danish and Japanese engineers comparing their recent feats. At the time, the Akashi Bridge was weeks away from completion, and symposium attendees took a tour. Donning hard hats, the Davenports climbed down into the under-

Figure 89. Serving on the French government's commission for the Millau Viaduct in the early 1990s, Davenport happily flooded his basement with box upon box of reports. PHB.cz (Richard Semik)/Shutterstock.com.

belly of the bridge to a public observation deck, and they traveled up, via the construction elevator, to the top of one of the 974-foot towers. "We climbed out at a dizzying height onto a small platform with a view reaching far over the inland sea," Sheila recounted in an article titled "Japanese Elevate Art of Bridge Building" for the *London Free Press*. "One end of the platform was without railings and I clung to the opposite side while our Japanese hosts described the finer details." The lore about the superstructure's finer details had traveled far and wide in the international media. It had the most formidable truss deck ever fabricated, 46 feet deep and 115 feet wide. Each of the two main cables, nearly four feet in diameter, was a bundle of 36,830 wires, which if outstretched would wrap around the earth seven and a half times. The only way to sling these massive cables from tower to tower was with a helicopter carrying a pilot lead wire that would support the catwalk used as a work platform to feed the cables across. And, perhaps most awe-inspiring, during construction in 1995 the bridge survived the Kobe earthquake, 7.2 on the Richter scale. A little over half a mile from the bridge, the epicenter of the quake cleaved open a previously unknown fault line, deep beneath the channel between the towers. Rigorous inspections revealed that the seismic event shifted the towers 2.6 feet further apart, and that one of the side spans was stretched about a foot, increasing the total length of the bridge by 3.6 feet and causing the midspan cables to hang four feet higher than specified. Since the suspenders had already been manufactured, the only fix was to modify the girder and truss system to match.

Equally impressive, from Davenport's perspective, were the lengths to which the Honshu-Shikoku Bridge Authority went with its wind program. Because of the bridge's colossal length, only section models would fit into the standard wind tunnel. But there were concerns, related to Akashi's total 12,828-foot distance, that testing only sections of the bridge wouldn't reflect the behavior of the structure as a whole. And so, for the world's longest suspension bridge the Honshu-Shikoku Bridge Authority commissioned the construction of the world's largest wind tunnel. The job was contracted to the Public Works Research Institute in Tsukuba City, a "science city" developed in the 1960s about 30 miles northeast of Tokyo. One of the senior

Figure 90. Japan's Akashi-Kaikyō Bridge, the world's longest suspension bridge, with a central span of 6,532 feet.

engineers on the project was Toshio Miyata, a postdoc at the Davenport Lab in the 1960s. Contemplating the colossal task before him, he sought out his mentor for advice at the Eighth International Conference on Wind Engineering, hosted by the University of Western Ontario in 1991. Davenport served as chair and made a dedicated effort to ensure attendance represented the field's increasingly international demographic (he insisted there be a budget item providing financial support for individuals who might not otherwise be able to afford to come, allowing, for example, his longtime friend and colleague Prem Krishna to make the trip from India).

Eight years hence, at the Kobe symposium and on the eve of the Akashi's inauguration, Miyata proudly described the end result of his labors, delivering a paper on the Akashi-Kaikyō's wind tunnel tests and the successful completion of the bridge itself. With the bridge's shore-to-shore distance of 2.4 miles, the wind tunnel scientists had decided a scale of 1:100 would suffice, making the model 131 feet in length. The dimensions of the wind tunnel came to 135 feet in width and 13 feet high, with a test section of 98 feet. The bridge was meticulously engineered with all the latest technology in computer simulations, but nonetheless, experimentation on the model in boundary layer conditions raised concerns about two serious wind issues. Tests showed that vortex shedding at the tower tops would send them vibrating apace, a problem that was quelled with the installation of ten 10-ton tuned mass dampers in each of the towers. Torsional flutter also proved to be a problem on the span, causing what Miyata described as "snake-like motions of a kind of wave propagation on the girder." Tailoring—designed and tested in the wind tunnel—made the structure more aerodynamically copacetic. Perforated gratings and curbs on the deck, and an 8-foot air gap between the deck and the truss, reduced the wind's clout (technically speaking, its torsional lift action). On a larger scale, a platelike stabilizer, a 6.5-foot fin, was attached under the median strip of the bridge deck to help equalize air pressures above and below. In the end, the structure was secured against winds of 206 miles per hour—a violent velocity unlikely to be unleashed by Mother Nature, even in Japan's typhoon-prone parts.

Like tall towers, as bridge spans grow, they enjoy aerodynamic benefits. For towers such as the Burj Khalifa, the wind forces decrease as the structure ascends into the higher altitudes, since the scale of the structure dwarfs the scale of the wind turbulence. Similarly, suspension bridges, as they get longer, are exposed to more variable conditions. Stretching as they do over the curve of the earth, the climate at one anchorage could be quite distinct from the climate at another, two or three miles away. The bridge's different sections experience different wind speeds, and this lack of correlation of wind forces works to the structure's advantage. While one gust causes excitation, another gust down the road cancels it out.

Italy's proposed Strait of Messina Bridge, running more than two miles, will link eastern Sicily with the Italian mainland. Dreams of such a crossing hark back as far as Roman times, when the bridge would have been fashioned from boats and barrels. Resurrected in varying forms in the eleventh, thirteenth, and nineteenth centuries, the project in its latest incarnation was vetted with some advance work at the lab. Was such a bridge even feasible, or was it complete folly? The design redoubles the daring of the Akashi Bridge: while Akashi's main span is 6,532 feet, Messina's will be nearly twice as long, 10,827 feet, suspended between two 1,255-foot-tall towers. As implausible and foolhardy as the enterprise might seem, Davenport never doubted for a moment that it was doable. He served as the wind expert among the leading engineers in the world who convened to form a review committee, assessing every aspect of the bridge's design over the course of a decade. He recommended and conducted a topographic model of the site, used to calibrate full-scale measurements of wind speeds previously made at two electrical towers nearby the bridge site, which for these purposes stood in for what would be the towers of the bridge itself, one sitting on the Eurasian tectonic plate, the other on the African plate, with a fault zone in the middle. Lots of wind speed measurements had been taken at these two towers but none for the span in between, where the bridge would actually be built.

The lab modeled the topography and measured the wind speeds along the bridge span for different wind directions, and then incorporated the previous full-scale measurements. In the end, Davenport concluded the bridge was viable, assuming certain issues were addressed with the highest order of care and consideration. He recommended a full-aeroelastic model study. But the length of the bridge

Figure 91. Davenport (left row, third from front) served as the wind expert among the world's leading engineers who convened to form a Strait of Messina Bridge review committee, assessing every aspect of the proposed design over the course of two decades commencing in the early 1990s. Courtesy of the Boundary Layer Wind Tunnel Laboratory.

Figure 92. In 1990 the lab ran extensive topographic tests on the proposed Strait of Messina Bridge, with one tower sitting on the Eurasian tectonic plate, the other on the African plate, and a fault zone between. Courtesy of the Boundary Layer Wind Tunnel Laboratory.

meant the lab literally could not fit it in. The Martin Jensen Wind Tunnel, built to test the Storebælt Bridge at the Danish Maritime Institute in Lyngby, took up the baton. Later, the Italians built their own wind tunnel at the Politecnico de Milano and conducted tests on a more advanced design. And when the Messina project went out to tender, each bidder assumed responsibility for the windworthiness of its creation. A Danish-Spanish-Japanese-Italian consortium called Eurolink won the competition, at which point all the elements of the design were wind checked yet again, and double-checked by peer review, with an international collaboration of facilities. Davenport's lab executed all the primary section model tests and double-checked tests on the towers. The results are proprietary and confidential, but Davenport was right. The bridge is feasible; it's just a matter of money.

For the Strait of Messina Bridge and its price tag of nearly U.S. $7.5 billion, the recurring roadblocks are the economic together with the political will. The project was set to break ground in 2006, when then Italian prime minister Silvio Berlusconi, one of the bridge's greatest enthusiasts, was defeated in a general election. The on-again, off-again plans for the bridge were cancelled by the new left-leaning government coalition, only to be resuscitated when Berlusconi was re-elected. Should it ultimately come into being, the Strait of Messina Bridge will break all the records, although there is precedent for such an ambitious and redoubling leap in bridge design, and one that brings the story full circle. In the 1930s, the George Washington Bridge's main span at 3,599 feet became the longest in the world, almost doubling the span of the reigning titleholder, the 1,850-foot Ambassador Bridge. The George Washington was also the bridge that emboldened engineers to go too far, too fast, without heeding the experience of history. But there hasn't been a major problem with a suspension bridge since the Tacoma Narrows disaster. As Davenport argued time and time again, it was a cautionary tale that bore repeating. The Strait of Messina Bridge has certainly attracted skepticism from critics at large, some falling to an easy cliché in proclaiming, "A bridge too far!" Heeding the wind, no bridge lover would suggest that such an archetypal structure—as Davenport would say, a means of connection and a measure of civilization, practical and symbolic both—could ever be a bridge too far.

IV

Project Storm Shelter

At the supper hour one Saturday in April 1996, the sky went a threatening shade of greenish black. Clouds, layer upon layer, low and heavy, scooted along the horizon in southern Ontario, offshoots twisting and swirling earthward. All day long, forecasters at the provincial weather center analyzed data and monitored computer prognostications. As evening fell, a thunderstorm took shape just northwest of Davenport's home base in London. The storm picked up force as it continued on its course, causing the severe weather specialists much consternation over whether or not to issue a warning—not wanting to gain a reputation for crying wolf, for issuing alerts too lightly or too often, lest they not be taken seriously. Finally, at 6:42 p.m., with confirmed reports of funnel clouds and even a twister touchdown, the weather warning bulletin went out:

A TORNADO HAS BEEN REPORTED AT ARTHUR WITH DAM-AGE.... THE STORM IS MOVING TO THE NORTHEAST AT [50 MPH] AND WILL AFFECT GRAND VALLEY AND SHELBURNE BY 7 PM. SEEK COVER IMMEDIATELY DUE TO THIS DANGEROUS STORM.

As it turned out, there was not one tornado but a tornado outbreak. The last outbreak in the region had occurred a decade before, when a series of thirteen twisters hit, the most severe an hour north of Toronto, killing eight and leaving eight hundred homeless— bisecting one home like a dollhouse, splitting it open to reveal a cross section of rooms, an iron and ironing board still standing. This time forecasters had spent an anxious few days tracking a destructive storm system barreling toward them along tornado alley. When the storm

Figure 93. After a 1996 tornado outbreak in Barrie, Ontario, Davenport documented the damage with composite photographs and copious notes. Courtesy of the Boundary Layer Wind Tunnel Laboratory.

finally hit, the tornadic perimeter extended from Owen Sound to Barrie, from Orangeville to Kincardine, cutting a path 235 miles long with numerous touchdowns and leaving insured damages totaling $7 million, but no fatalities.

Davenport had never been a storm chaser per se, although once, when the lab's Bjarni Tryggvason, an avid pilot, suggested they rent a plane for a flyover of some tornado tracks, the lab's director could hardly decline. And on this occasion, when he heard a tornado had touched down 100 miles as the crow flies from his home, he understandably wanted to take a look. The next day he and Sheila set out on a Sunday drive with an ulterior motive. Heading north on the highway, they encountered a fleet of utility trucks. Davenport correctly surmised they were en route to fix power lines downed by the tornadoes, and they led him straight to the point of touchdown. Access to observe the damage on site wasn't quite so easy to come by. When Davenport told the utility crew he was a wind engineer, they didn't budge. Not one to take no for an answer, he turned the car around, found a back-road route through a farm, and convinced the farmer to let them traipse through his field. There Davenport found the tornado tracks he was after, and he documented the evidence with photographs and copious notes.

When approached with such a scientific investigative eye, disasters can have unintended positive consequences. Surveying destruction for causative clues can reveal what could have and should have been done in the way of prevention. Prevention had always been at the top of Davenport's mind—and not just preventing a crisis from befalling a single bridge or skyscraper. "It's almost an epidemiological approach," he noted of the lab's work over the years. "You understand diseases through not just one sick patient but through groups of patients. At the lab we take in patients, as it were, and we can generalize on the particulars based on these individual structures. We've tested 100 very tall buildings, and from that experience we now have a good idea of what the general characteristics are of this group of structures. That is knowledge we could only get at piece by piece." Davenport viewed this accumulated wisdom as a rich mine, a repository that, if put to good use, serves the public interest. Factoring in hurricane winds during tests on the World Trade Center or the Sunshine Skyway Bridge in turn could inform broader research aimed at preventing damage caused by hurricanes everywhere. The next step would then be to incorporate what was learned into industry's best practices, into codes and regulations, thus ensuring that the knowledge benefited society as a whole.

With the lab steady on its course dealing with one structural patient after another, disaster prevention rose to the top of Davenport's list of priorities during the final act of his career. His timing proved propitious, as the 1990s were declared the International Decade for Natural Disaster Reduction by the United Nations. The special decade was the brainchild of an eminent American geophysicist, Frank Press, president of the U.S. National Academy of Sciences. Press had first suggested the idea in 1984 during his keynote address at the Eighth World Conference on Earthquake Engineering. He argued that humanity was increasingly vulnerable to natural disasters as urban populations exploded, ever increasing the stakes. This, com-

Figure 94. Inducted as a Fellow of the Royal Academy of Engineering in November 1987, Davenport and the Duke of Edinburgh discussed a violent 100-mile-per-hour windstorm that one month earlier had devastated London's parks. Courtesy of Sheila Davenport.

bined with more settlement in high-risk areas such as along coastlines (out of either necessity or indulgence) and more extreme weather events due to climate change, as well as the heightened complexity of modern life and dependence on infrastructure, all added to the urgency, and on an international scale. "This special initiative would see all nations joining forces to reduce the consequences of natural hazards," Press said of his plan for the dedicated decade. Early in the planning he joined forces with Davenport, tapping him to lead the Canadian National Committee. Davenport took the proposal to the Royal Society of Canada and the Canadian Academy of Engineering, and with their backing Canada became an early catalyst and sponsor as the project was presented to the United Nations. In 1989, UN Resolution 236 announced that the International Decade for Natural Disaster Reduction would commence in 1990.

The mandate was clear: by the turn of the twenty-first century, to reduce the loss of life, property damage, and social and economic disruption caused by natural disasters—tornadoes, earthquakes, tsunamis, and other calamities of nature. In the twenty years prior, natural disasters had killed three million people, adversely affected 800 million lives, and caused $300 billion in damages. By bringing to bear scientific and engineering know-how, the goal of the International Decade for Natural Disaster Reduction was to temper these deleterious effects of natural disasters, or rather natural hazards. Part of the prescribed paradigm shift required a rethinking of terminology. A "disaster" is an event that overwhelms a community. It does not necessarily mean a lot of people die, or that the bill for damages is astronomical. But it means a community, whatever its capacity, is unable to cope. The term "natural disaster," however, is misleading. It is a severe natural event that has disastrous consequences, a natural event that turns into a disaster. As dangerous as these calamities are, there is a common misconception that destruction and loss are part and parcel. As Davenport noted in one of his reports during the International Decade, "History and mythology are filled with tales of the dread humans have for catastrophes. The fatalistic 'they are inevitable' response to these fears is an ancient one, but today we know that by taking prudent reasoned actions, we can often prevent disasters and mitigate their impacts." Damage from natural hazards, especially the loss of life and property, could be viewed as collateral damage: to a large part avoidable, preventable, unnecessary.

Davenport had long taken a proactive stance in guarding against not only the so-called well-behaved wind and weather but also more severe and hazardous climatic phenomena. Perhaps his earliest contribution to the disaster mitigation mandate dated to the 1960s, when he developed the world's first statistically based seismic zoning map, working with Pacific Geoscience Centre scientist W. G. Milne, then a PhD candidate. Their study focused on the regions of maximum earthquake activity—the St. Lawrence Valley and the Pacific coastal

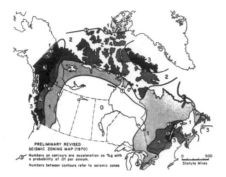

Figure 95. The seismic zoning map for Canada produced by Davenport and W. G. Milne. The graph is keyed to the peak ground acceleration levels expected at 1-in-100-year probability level and shows the areas of greatest hazard (the darkest shaded) along British Columbia's west coast and in the eastern region along the St. Lawrence River. Courtesy of the Boundary Layer Wind Tunnel Laboratory.

belt of British Columbia. It was based on calculated peak ground accelerations for the earthquake catalogues held by the Dominion Observatory in Ottawa. Davenport and Milne performed statistical analyses of the data and created geographic contour maps detailing expected peak acceleration amplitude at a probability of 1/100 per year. The value of this analysis, dubbed the "Davenport-Milne method," did not go unnoticed. It was implemented in the 1970 edition of the National Building Code of Canada, replacing a more simplistic map that ranked regions by their seismic activity but failed to factor in the statistical nature of regional risk.

Davenport also turned his statistical microscope on the risk of tropical cyclones—hurricanes and typhoons, the most devastating of all natural phenomena. These were not of utmost concern in his own backyard in southern Ontario, or even in North America, but he was always adamant that the knowledge gleaned and the applications developed in service of prevention reached across all boundaries and borders. And, most important, that they should reach vulnerable populations in the developing world. "Some of the most populated areas in the world are affected by tropical cyclones including countries whose urban populations are doubling every 7 to 10 years," Davenport noted in a paper presented at the World Meteorological Organization's 1984 conference on urban climatology in Mexico City. "The evaluation of the risks imposed by tropical cyclones thus is of critical importance to these countries, offering the potential to greatly reduce the loss of human life and the damage caused by these storms. Within particular developing countries assessing the vulnerability of different regions to tropical cyclones is an important first step toward the longer-term goal of natural disaster prevention involving land-use and zoning laws to control development in areas where the risk is high." Believing that Canada's tradition of humanitarian aid extended to mitigating natural disasters worldwide, he worried about communities where concern for preparedness in surviving disasters was often and understandably neglected, trumped by concern for subsistence and day-to-day survival. To this end, he didn't focus his attentions on the costly disaster prevention solutions that only the United States, Canada, and other developed countries could afford. He was interested in solutions that were simple and inexpensive. He explored how subtle changes in roof design could save lives during a hurricane. For instance, putting a rope over a roof and tying it down is a simple and cheap tactic, though one that some engineers, motivated by fat contracts or lucrative patents, would not be interested in exploring. With the hurricane wind risk studies, the focus first went to familiar Canadian and American regions, with the study subsequently expanded to Hong Kong and Australia. Ulti-

Figure 96. Davenport lecturing at the Hong Kong University of Science and Technology, 1988. Courtesy of Sheila Davenport.

mately and more critically, Davenport conducted the most detailed case studies investigating the countries of Belize, Jamaica, and Mauritius and the entire region of the Caribbean extending from Trinidad to Puerto Rico—home to developing societies in need of support in generating resources and knowledge about prevention and mitigation.

The most basic requirement in mitigating disaster from tropical windstorms is knowing what one is up against, which means having reliable information on wind speeds. Obtaining wind speed predictions for tropical storms, however, is more difficult than assessing well-behaved wind. For the latter, reports from airports and strategically placed anemometers atop high buildings suffice for data gathering. These traditional approaches don't work with tropical cyclones since anatomically such storms are relatively small and hard to capture: a storm being a couple hundred miles in diameter, with the highest winds at the center, the swath of interest encompasses only about sixty miles. Storms at any particular locality are also difficult to capture because of their infrequency, making the placement of anemometers problematic. And even if instrumentation were fortuitously placed, flying debris or loss of power during the storm would likely render it dysfunctional. All of which is to say that the traditional data-gathering methods can't be counted on.

Davenport and his team overcame these obstacles by predicting cyclone wind speeds using the Monte Carlo method. The method had been invented, and the term coined, in the 1940s by physicists Stanislaw Ulam, Enrico Fermi, and John von Neumann, who were running nuclear weapons simulations at Los Alamos. Ulam's uncle was an avid gambler at the Monte Carlo Casino in Monaco, and the random and repetitive nature of gambling seemed an apt analogy for this computational investigation into the infinite stats quantifying the physical world—the computer churning out results of various scenarios using repeated random samplings from vast amounts of data. Although it had been applied in meteorology to investigate marine storm surges, the Monte Carlo method had never been used to simulate storm winds. Davenport first tried it at the lab to test specific building sites, such as the Pam American Building in New Orleans, with the computer program written by Tryggvason, who was well-suited to the task, having done a stint as a meteorologist with the cloud physics group at the Atmospheric Environment Service in Toronto before joining the lab. The core of the program was a wind field model that could realistically generate the winds occurring for a storm with known large-scale parameters such as a storm's central barometric pressure, the radius and pattern of maximum winds, and variables that describe the geographic distribution of these storms, their direction of movement or tracks, as well as their annual occurrence rate. The wind field model could be verified by comparison with a handful of storms that had left good wind records of their passing. The model proved to be a good mimic. These large-scale parameters were statistically stable over large geographic regions and could be much more easily acquired since they varied more gradually over a storm's life span, often lasting a week or more as the storm tracked across land and sea. And since these metrics could be readily sampled, rich historical records existed. In particular, the National Hurricane Center in Miami, Florida, held a stash of HURDAT (as in hurricane data) tapes documenting all known North Atlantic tropical cyclones from 1886 to 1983. With this cache at its disposal, the lab's computer simulations randomly sampled storm specs and generated unique profiles for storms that would occur at any given location over a great many years—ten thousand years, or more if desired (at a minimum, the simulation would be run repeatedly until the results stabilized statistically).

As Davenport explained, "In essence this so-called 'Monte-Carlo' method creates by computer simulation a large number of time histories of tropical cyclone passages past a given locality." At any particular location, he noted, "the historical data records are analyzed to produce a statistical representation of the tropical cyclone parameters required by the wind field model used in the simulation. Tropical cyclones are then computer-generated, moved past the site of interest within some specified distance"—past the Bronx-Whitestone Bridge, for instance. The diameter of the typical simulation circle was 150 to 300 miles, and the model was capable of adjusting once a storm made landfall, as well as making modifications to account for change in terrain roughness, whether in New York or New Orleans. During the course of the storm simulations, the computer-generated wind speeds and wind directions at the site of interest are recorded. Since the storm specs created during the simulation bear the same characteristics as the storms that passed through historically, each new time history of wind speeds generated by the computer could be taken as an accurate estimate of a storm that might pass by. Cumulatively, an accurate estimation of storm wind speeds at the site of interest could be generated for a 10,000-year period, and this in turn would allow the extraction of an accurate design wind speed for that site.

Repurposing this program for the broader regional climes of the Caribbean, Davenport zeroed in on an island chain stretching from Jamaica to Trinidad and Tobago. The HURDAT repository provided statistics on fourteen hurricanes over the last century that had passed within about 30 miles of Antigua, among the Leeward Islands in the Caribbean Sea. With the Monte Carlo hurricane simulation's design wind speeds in hand—the winds ranging from 68 miles per hour in Trinidad to 93 miles per hour in Dominica, at a height of 33 feet over open water—another study then zoomed in further, examining in close-up the influence of island topography. Scrutinizing the island of Nevis, with its impressive volcanic peak but otherwise compact and simple topography, Davenport constructed a 1:3,000 scale model and tested it in the boundary layer wind tunnel. Not unexpectedly, the results showed that varying topography would produce significant increases and decreases in wind speed. Traditionally, settlements in the Caribbean often took advantage of natural shelter from surrounding hills, while structures on hilltops on a windward coast expected to endure winds even stronger than if they were over open water. With the combined results of the two studies, Davenport generated a microzoned hurricane risk map for the entire Caribbean region. It was all part of a very specific task, for Davenport had been retained to write the wind loads section of the Caribbean Uniform Building Code (CUBiC), part of the larger Pan-Caribbean Disaster Prevention and Preparedness Project. And following Hurricane Gilbert in 1988, Davenport zoomed in further again when he undertook diagnostic surveys of the damaged buildings in Jamaica. Focusing on the damage to two structures, a pipe-supported canopy at Princess Margaret Hospital in Morant Bay and a series of lamp poles

Figure 97. Wind tests on the island of Nevis on a 1:3,000 scale model, as well as other tests in the region, allowed Davenport to generate a microzoned hurricane risk map for the Caribbean. Courtesy of the Boundary Layer Wind Tunnel Laboratory.

on the road leading to the campus of the University of the West Indies, he reached a reassuring conclusion. The failure wind speeds of those structures were within the range prescribed by CUBiC, which had been published three years before Gilbert, but long after the canopy and lamp poles had been built.

Contemporaneously with his work in the Caribbean, Davenport, the consummate multitasker, worked on a loads and forces code for the Swiss-based International Organization of Standardization. An ISO technical committee legend has it that in a scant few hours in Vienna he drafted an entirely new and minimalistic code of practice for the wind loads on buildings, which immediately became the basis for the Swiss Code and eventually the Eurocode. One colleague noted of his contribution to that document, "Dr. Davenport's work eschewed dogmatism about the approach to providing safety, recognizing that there may be several adequate methods for design."

Figure 98. After Hurricane Gilbert in 1988, Davenport conducted diagnostic surveys of damage in Jamaica. Courtesy of the Boundary Layer Wind Tunnel Laboratory.

The International Decade for Natural Disaster Reduction sought safety in numbers, with concerted efforts toward prevention and preparedness on an international scale. Some 140 countries established national committees or official programs, bringing together representatives from government, academia, industry, volunteer organizations, and NGOs. By the year 2000 all countries were to have in place comprehensive assessments of their vulnerabilities and risks, plans for prevention and preparedness, and access to global, regional, national, and local warning systems. The mandate seemed straightforward enough. Even with these basic targets, however, the huge size of the task made it difficult to get any purchase on progress. The Canadian National Committee surveyed the scientific literature and estimated that an earthquake in Vancouver, British Columbia, one of the nation's densely populated hotspots, could cause $32 billion in damages. It warned that neither the economy nor the insurance industry of a developed country could absorb the losses without severe dislocation, and much less a double disaster such as hit Japan in 2011, when the most powerful earthquake in recorded history triggered a deadly tsunami and caused damages exceeding $300 billion. In developing regions, a single natural disaster could derail the entire economy, easily exceeding a nation's GDP, diverting development funds to disaster recovery, and setting society back for generations to come. But yet, during the International Decade, any substantive action toward prevention and preparedness proved elusive.

There was some piecemeal progress. Following the example of Davenport's seismic zoning map, Emergency Preparedness Canada produced the "North American Natural Hazards and Disasters Map," mapping the vulnerabilities of Canada, the United States, and Mexico. *National Geographic*, the time-honored arbiter of humankind's relations with the natural world, took notice, publishing a glossy foldout version in its July 1998 issue. But the bigger plans more often than not fell short. Plans articulated by Davenport and the Canadian National Committee that never came to pass included a center for the engineering investigation of natural disasters—a command central that would coordinate studies of wreckage from natural disasters immediately after the fact. "A double tragedy following natural disasters is that frequently vital technical information is cleared away before it can be assessed and documented," Davenport noted in the inaugural issue of *Natural Hazards Review*. "The bulldozing of transmission towers, for example, has happened following both tornado and ice storm damages. However, if properly investigated, natural disasters can provide key information to plan and design new built-up environments, evaluate and upgrade existing built-up environments, and better assess risks from an insurance point of view." This plan for

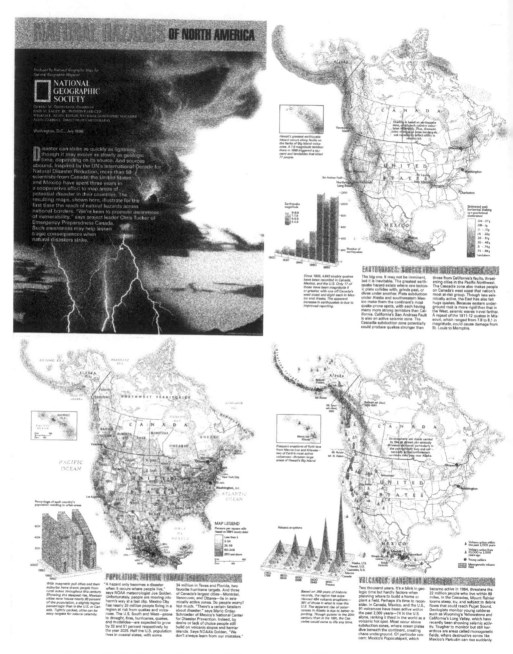

Figure 99. The Natural Hazards of North America foldout map series, published in the *National Geographic*, July 1998. NG Maps/National Geographic Stock.

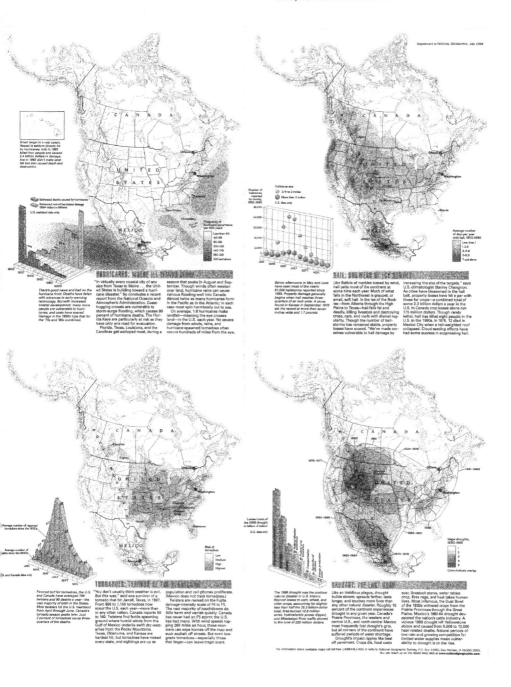

an investigative center languished, despite encouragement and support from Emergency Preparedness Canada and the Insurance Bureau of Canada. Ultimately, on political, economic, and civic levels; it lacked the collective will.

The symbolic if not substantive success of the International Decade could be seen in the UN's declaration of the second Wednesday in October as an annual International Day for Natural Disaster Reduction. In 1995 the vaunted NPR program *All Things Considered* ran a feature. "The United Nations is observing the day to highlight how human activity has magnified nature's cruelest blows," host Robert Siegel said in his introduction. "Where once an earthquake, a tornado, a hurricane or a flood would leave in its wake a measure of devastation it now tends to destroy more property and claim a greater number of lives. . . . When it comes to natural disasters, nature makes humans the victims," he concluded, "but humans also must accept responsibility for being the culprits." At that, the day, like the decade, was a successful medium for spreading awareness. Yet nearing the turn of the twenty-first century, UN secretary-general Kofi Annan took a somewhat somber look back. "We confront a paradox," Annan said. "Despite a decade of dedicated and creative effort by IDNDR and its collaborators, the number and cost of natural disasters continue to rise. The cost of weather-related disasters in 1998 alone exceeded the cost of all such disasters in the whole of the 1980s. Tens of thousands of mostly poor people have died. Tens of millions have been temporarily or permanently displaced. 1998 was, in fact, a truly disastrous year." He itemized hurricanes George and Mitch— the deadliest Atlantic storm in 200 years—which killed more than 13,000 in the Caribbean; a cyclone in India that killed 10,000; the thousands that died and millions more who were displaced in China's Yangtze flood; the earthquakes in Afghanistan that killed more than 9,000; the fires in Brazil, Indonesia, and Siberia that ravaged forests. "The cost of disasters in the 1990s was some nine times higher than in the 1960s," he noted, "and it is becoming increasingly clear that the term 'natural' for such events is a misnomer. No doubt there will always be genuinely natural hazards—whether floods, droughts, storms or earthquakes. But today's disasters are sometimes man-made, and nearly always exacerbated by human action—or inaction." Early in the decade there had been the so-called Storm of the

Century, a massive cyclonic event in March 1993 that sprawled from Canada to Central America. It might have been enough to mobilize some forward-looking and long-term preemptive action. But as soon as we emerged from the trauma of reconstruction and recovery, we forgot the lessons to be learned. Similarly, in Japan, cautionary tales went unheeded. Hundreds of engraved tsunami stones, some more than six centuries old, dotted the coast, issuing a warning to villagers from their ancestors: "Do not build your homes below this point!" Other mammals, notably elders among elephant populations, have been said to possess disaster management skills superior to humans'. During drought, elephant matriarchs draw on memories of hard times several decades past and shepherd kin to faraway watering holes they frequented in their youth. Human disaster memory, by contrast, along with our attention spans, only seems to get shorter, even as we become more scientifically prescient. Scientists are analogous to our elders, but they find it nearly impossible to get their message through to the rest of the herd.

The antidisaster decade had started with great fanfare, with great optimism and promise, but the momentum fizzled. "Society is not good at dealing with problems that have low probabilities of occurring, but high consequences when they do," noted Davenport. "When engineers deal with the possibility of severe wind storms or earthquakes, they make decisions based on risk assessment. Where communication breakdown often occurs is in dealing with low probability but potentially disastrous results." And as crass and cynical as such an observation may be, responding to a disaster is a better photo-op for a politician than preventing disaster, says Gordon Mc-Bean, a climatologist at the University of Western and a longtime friend and colleague of Davenport's who has taken up the disaster risk and prevention mantle. "If you do it right, nothing happens," says McBean. "The politicians can't stand up and say, 'See, I did all this, and nothing happened.' That usually doesn't get much attention. Hurricane Katrina is a tragic example where the advice of the scientific community was ignored. The scientific community had long described in great detail—including the number of body bags that would be needed—what would happen when a big hurricane came through New Orleans. The choice of the United States government was to give the engineers 10 cents on the dollar of the budget pro-

Figure 100. The lab conducted the first wind studies on communication towers and transmission lines, developing experimental and analytical techniques that are now common practice in assessing these crucial components of infrastructure. Walter Matheson/Shutterstock.com

posed to fix the levees. They were told, 'That's enough. We've got a war in Iraq to fight.'" Canada was hamstrung by the essentially bankrupt state of federal finances and a government interested only in funding research for new electronics gadgets that would create jobs and spawn spinoff companies. That the government should fund research for public policy or public good, investigating questions such as how do we design better building codes? or how do we understand the climate and the causes of extreme weather events?—that was a nonstarter. The philosophy of investing for long-term benefit is a near extinct axiom of public logic.

All didn't come entirely to naught. As the 1990s came to a close, a permanent international body assumed the International Decade's mandate, in the form of the UN International Strategy for Disaster Reduction. McBean sits on the Science and Technology Committee and serves as chair of the Integrated Research on Disaster Risk scientific program, which is under the International Council for Science and International Social Sciences Council—all evidence, at least, that the efforts continue. Davenport himself was only somewhat slowed in his efforts by the onset of Parkinson's disease. And his formidable passion and perseverance continued as a catalyst, propelling seeds sown during the International Decade to germinate and produce fruit.

"Ingrained in Alan was the belief that you can build safely," recalls Paul Kovacs, a Canadian National Committee member from the private sector (at the time vice-president of policy development at the Insurance Bureau of Canada). "You can build longer bridges and taller buildings, and these do not have to be dangerous undertakings. All the structures you build can be perfectly safe, if you have done your homework." Working with Davenport during the International Decade, Kovacs came to see the need to unite scientists, and all they knew about prevention, with insurers and their vested interests. He rallied insurance companies to support his vision for a permanent antidisaster think tank, and the Institute for Catastrophic Loss Reduction came into being in 1997, funded by member insurers such as Lloyd's of London. Davenport served as the institute's research director and forged a partnership with his lab at the University of Western Ontario. And under the auspices of the institute, Davenport witnessed the realization of several of the goals left unfulfilled during the International Decade.

Drawing attention to low buildings, for example, was one of Davenport's long-standing concerns. Low buildings typically struggle to compete with statuesque skyscrapers and elegant bridges in capturing engineers' attention—paradoxically so, since the severity of a natural disaster's devastation, the ability of a community to cope and recover, turns on the survival or failure of these Everyman structures—houses, multiresidential dwellings, hospitals, schools, churches, and community centers, as well as infrastructure hubs for communication towers, transmission lines, and cooling towers. During Hurricane Katrina an estimated 300,000 homes were destroyed or damaged, primarily by flooding. But low buildings, for all their importance, are typically the forgotten stepsisters of structures.

One of the key issues influencing a low structure's fragility is quality control in construction, or lack thereof. A lack of quality can stem from any number of deficiencies that collectively lead to failure. Most fundamentally, these are one-off structures, meaning that designs are rarely engineered, let alone run through a wind tunnel, there being few incentives for costly research and many incentives for cutting costs. The root of the problem can originate in any number of places: in the design stage, with an inappropriate structural system or unforeseen weak links in the load paths; in the materials stage, owing to poor specifications or the substitution of components; in the construction stage, as a result of lowest-bid practices, poor workmanship, or neglected inspections. Generally speaking, low buildings often manage to avoid conforming to engineered design. And many an engineer or policy wonk would point out that the nebulous system of low-rise building codes often seems unduly influenced by the pressures of lobby groups, catering to the competing interests of various sectors within the construction industry—an industry perceived as "dangerous and dirty" and that, unlike the automotive, food, and pharmaceutical industries, is sorely lacking for regulation.

Davenport witnessed these dynamics firsthand when the Metal Building Manufacturers Association hired him to provide some scientific backup for their lobbying efforts. In 1972 the MBMA was faced with a new building code in the United States that was punitive for low-rise buildings as far as wind loads were concerned. Because metal buildings are much more sensitive to wind loads than concrete buildings, the Metal Builders, representing manufacturers of prefab

Figure 101. A 1972 visit to the lab by the Metal Building Manufacturers Association, which hired Davenport to provide some scientific backup to its lobbying efforts to improve building code wind load requirements for low buildings. Courtesy of the Boundary Layer Wind Tunnel Laboratory.

metal structures and the like, wanted to assess the validity of these wind loads. Just the year before, Davenport had executed a similar study on low-rise buildings in Denmark (not known for its skyscrapers), commissioned by Martin Jensen, he who first asserted the importance of the boundary layer. Backed by the Danish Government Fund for Scientific and Industrial Research, Jensen was interested in seeing the effects of unsteady wind on some basic low-building shapes, simple square and rectangular plans. The focus went to roofs, where negative pressure or suction generated by wind loads—causing corners and windward edges to lift off—was found to be of greater concern than positive pressures. The Metal Builders' study took the testing on low buildings a few steps further, introducing a number of novel techniques such as determining loads through influence lines, and pneumatic averaging—the latter technique conceived by Dave Surry, then a research director, and grad student Ted Stathopoulos. (Known for his collegial collaborative style, Davenport was always one to give credit where credit was due; see the sidebars, "Averaging the Air," page 204, and "Raising the Roof on Low Building Testing," page 206). As with the lab's approach to bridges, these techniques measured average loads over areas, like turbulence in a rough sea, rather than just point loads. "If you just measure at one point," explains Surry, "you will certainly see very high peaks, as the worst case wind blows over the roof. But if you base codes on these peaks you are overestimating. If you take an area, averaging the pressures at every instant, you find that as you average it out over a bigger and bigger area, it gets smoother and smoother. Little rowboats get tossed around in the waves," he adds by way of an analogy, "whereas with an ocean liner the same waves don't have the same impact." Using these new techniques Davenport measured the way load varied and spread over the tributary area—loads congregating on a beam much as tributaries feed a river. Ultimately, the lab's work on low building codes not only improved the Danish design and helped the Metal Building Manufacturers argue their case, it also advanced codes for low buildings both locally and internationally. Again, the big-picture goal was always in the back of Davenport's mind. He wanted individual studies to make a larger contribution. He always had his eye on how seeds for generalization could be planted with each special case.

Averaging the Air

Necessity was the mother of invention that inspired the lab's David Surry and Ted Stathopoulos to conceive of the pneumatic averager in the 1970s. They needed to capture more accurate measurements of pressures on low building roofs generated by the gustiness of the wind and the turbulence of separated flow (the technique was later applied to other structures, such as the Sunshine Skyway Bridge). "As we learned in high school," explains Stathopoulos, professor of civil engineering at Concordia University, "pressure on a surface is defined as force divided by area. We usually think of pressure as a constant—if there were a perfectly steady wind blowing towards a building façade, the mean force divided by area would give the mean pressure, and that would be the end of it. But as we all know, the wind is gusty. Wind is particles of air moving at different speeds, and so particles hitting the surface of a structure at each instant hit it at different speeds at every location. There is a maximum speed and a minimum speed, and naturally the total dynamic pressure on the surface area is lower than that of the fastest-moving particle. The question is, what is the reduction?"

This reduction, of course, would allow for economy of design; like towers and bridges, low buildings need not be excessively conservative in their design. In an attempt to measure the reduction, Stathopoulos and Surry invented the pneumatic averager—the word "pneumatic" coming from the Greek *pnein*, "to blow, to breathe." They needed a way to more precisely measure the instantaneous average wind-induced pressure. At the time, the lab's data acquisition system could handle a mere sixteen channels simultaneously; investigators could measure the pressure based on a mere sixteen taps from the model feeding data into the computer. With the pneumatic averager, however, investigators could measure pressure based on hundreds of taps. For instance, seven input taps (see figure 102), or even as many as twenty taps, measured the pressure on the model. The array of taps then converged at a middle chamber, where the air mingled and achieved equilibrium. Finally, that physical averaging effect funneled into a central tube, providing the output for only one of the sixteen data acquisition channels.

Based on this and other studies on low buildings, investigating different terrains and employing the advanced techniques of pneumatic averaging as well as influence line measurement, Davenport proposed what Stathopoulos describes as the "famous and controversial 0.8 factor." In the Canadian code, wind pressure coefficients for low buildings came to be formulated by considering the most critical or highest measured pressure values factored by 0.8, or reduced by 20 percent. "This factor generated one of the major controversies in the history of wind engineering," says Stathopoulos. The American wind engineering community did not want to apply the 0.8 factor, since it was far from conservative. And true enough, Davenport had devised this factor mainly from intuition, having limited experimental data with which to work. "Not all engineering problems can be solved precisely," says Stathopoulos. "But Davenport had what I call an engineering feeling for things. He knew how things would work.

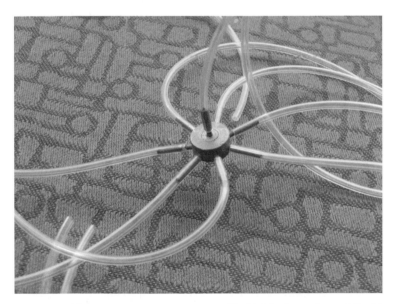

Figure 102. The pneumatic averager invented by Ted Stathopoulos and David Surry. Courtesy of the Boundary Layer Wind Tunnel Laboratory.

He could feel the structures and how they would respond. He could provide answers even before doing experiments." As far as the legitimacy of the 0.8 factor went, a few years later, experimental work by another PhD student, Eric Ho (now a director of the lab), confirmed and justified Davenport's radical intuitive approach.

Raising the Roof on Low-Building Testing

Low buildings, for all their importance in providing basic shelter, especially during natural disasters, rarely attract much engineering attention. In the late 1970s, however, the lab's Ted Stathopoulos, now at Concordia University, developed a technique for defining wind loads on low-rise buildings in groundbreaking doctoral thesis experiments. He equipped the model buildings with two instrumented bays that could be moved relative to one another, each with eight rows of four pressure taps on the roof, and five pressure taps on the front and rear walls. The characteristics of the pressure at each of these taps was not only measured with a conventional experiment, but also with a second experiment that measured characteristics of spatial averages of the pressures—that is, the pressures important to larger areas, for which point pressure results would be conservative (see figure 103). To do this, the roof and wall taps were pneumatically averaged using the devices invented by Surry and Stathopoulos so as to provide their instantaneous spatial average. This yielded ten instantaneous loads acting on areas of the structure that were substantially bigger than those represented by single taps, effectively reducing the largest peak pressures compared to those on individual contributing taps, and taking out the highest-frequency components. Furthermore, the computer instantaneously formed combinations of the ten component roof and wall loads. In this way, not only could spatial averages over still larger areas be obtained, but these loads could also be combined with hypothetical structural characteristics of the bay—to calculate overall loads, deflections, and stresses caused by the instantaneous load distributions. As a result, still more of the highest-frequency components were happily lost. Later, simple load distributions were derived analytically to recreate the effects observed, which in turn could form the basis of simple code models of all aspects of the loading, from the local peak pressures at points, important for fasteners and cladding, to the loads that were important for the design of the entire structure.

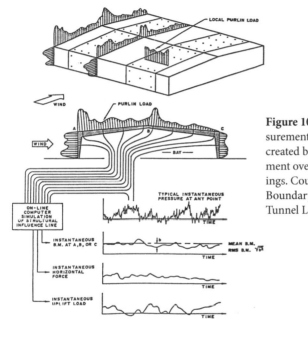

Figure 103. The measurement of pressure created by wind movement over low buildings. Courtesy of the Boundary Layer Wind Tunnel Laboratory.

All the same, any truly pervasive and dramatic change—on a scale that would mitigate the billions of dollars of damage sustained during hurricanes or earthquakes—demanded nothing short of a paradigm shift embracing a simple guiding principle Davenport called "total quality management." This necessitated a macro-scale integration of interests, the interests of all stakeholders—tenants and owners, investors and bankers, builders, architects, engineers, code writers, building inspectors, legislators, and insurers. The role of insurers, Davenport felt, could be particularly powerful and far-reaching. "It buffers the risk of all the stakeholders; it can provide incentives for improved design through adjustment of premiums; it can be an effective influence in setting building standards. There are strong arguments," he noted, "for expanding the role of insurance in the management of catastrophes." The Institute for Catastrophic Loss Reduction, then, Davenport saw as a welcome leap forward.

And the institute soon paid dividends. It played a crucial role in bringing to life a variation on a project Davenport had set his sights on in the 1990s called Project Storm Shelter. He envisioned and proposed an international venture whereby engineering scientists would assess the vulnerability of housing in windstorms and earthquakes in both developed and developing countries. The project would involve a comprehensive series of wind tunnel tests studying the geometry of basic low building structures, providing a systematic classification of their performance and recommendations for improving performance to achieve more "disaster-proof" construction. "What is needed," said Davenport, ever the proponent of comprehensive simplicity, "is a catalogue of pressure coefficients for a wide variety of

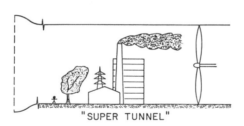

Figure 104A. Davenport imagined a "supertunnel" that would marry full-scale monitoring with full-scale testing. Courtesy of the Boundary Layer Wind Tunnel Laboratory.

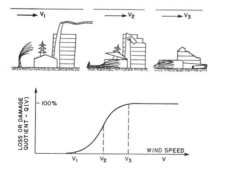

Figure 104B. Davenport's illustration showing his concept of the "damage quotient." Courtesy of the Boundary Layer Wind Tunnel Laboratory.

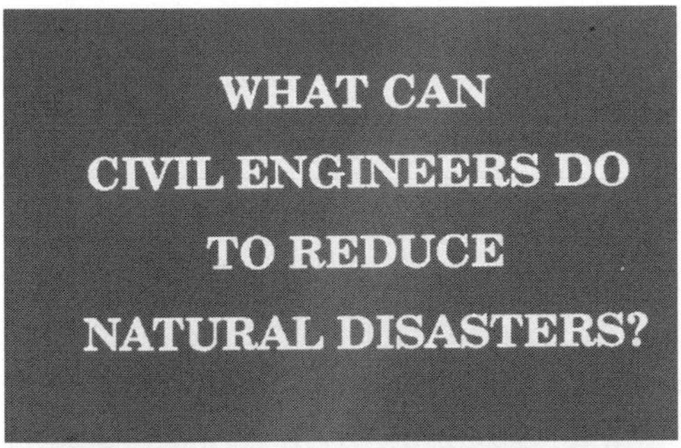

WHAT CAN CIVIL ENGINEERS DO TO REDUCE NATURAL DISASTERS?

WHAT CAN STRUCTURAL ENGINEERS DO TO REDUCE DISASTERS?

Figures 105A–B. The United Nations declared the 1990s the International Decade for Natural Disaster Reduction, occasioning many pointed talks on the subject by Davenport. Courtesy of the Boundary Layer Wind Tunnel Laboratory.

basic shapes, with a default method for establishing pressures on shapes not included." Backed by the International Association of Wind Engineering, a multinational steering committee launched the project in preliminary form in 1998, but without sufficient funding the initiative stalled.

An extension of Davenport's Project Storm Shelter included a presumably specious if hypothetical "supertunnel," which would marry full-scale monitoring with full-scale testing. Davenport imagined a wind tunnel that could accommodate tests not only on full-scale low buildings but on entire full-scale neighborhoods. As he once described, "To clarify the thinking on the question of loss predictions, I would like to carry out a 'thought experiment' involving a hypothetical wind tunnel of enormous size in which entire full-scale buildings and city blocks can be 'tested'. The advantage of testing full-scale structures in their real setting is that their failure mechanisms and loading can be monitored exactly as it would happen, without any speculation." Davenport provided hand-drawn mock-ups of a grouping of buildings in such a wind tunnel, showing how a gradual increase in wind speed produced a cascade of mounting distress, minor damage, more serious failures, and total destruction. He figured that if each of these "damage states" was translated into a corresponding "loss equivalent," engineers could then estimate the fraction of the total loss, or the "damage quotient," produced at each wind speed. This practically minded prediction of loss, he argued, was crucial in understanding how to strengthen structures and reduce damage.

Though just a thought experiment, the fanciful supertunnel turned out to be not so far-fetched, and not so far off in the future as a bona fide scientific experiment. In 2001, the institute was keen to make its name with a big-impact project. Davenport, meanwhile, was still keen on doing a massive and traditional wind tunnel study of homes to amass a credible database of stats. The team at the lab sat around the boardroom table and brainstormed some possibilities. Surry argued that a nontraditional approach might have greater effect. Investigating, for instance, how full-scale houses behaved when subjected to design winds—that research might be of more consequence in homing in on the problems plaguing low buildings. He had in mind something along similar lines of a project conducted at the lab not long before by Davenport and Mike Bartlett, a professor of civil and environmental engineering at the university. The duo had tested designs for a high-performance, low-cost prefabricated house made of corrugated cardboard that could be erected quickly for temporary or short-term use in post-disaster settings, when the existing housing stock failed. Manufactured by Durakit, these shelters—one called the Disaster Relief Unit, assembled in hours, and another the Instant House, assembled in a few days—subsequently won a U.S. $13.1 million contract with the U.S. Federal Emergency Management Agency, providing 350 shelters during the aftermath of hurricanes Katrina and Rita. Bartlett had conducted a full-scale test on one of these houses.

This in turn prompted Surry to suggest that the lab build an inexpensive traditional house and test it to destruction. With a grant of $100,000 to play with, they couldn't initially afford to build a house and blow it down, but they did put together a convincing feasibility study, and by the summer of 2007 an incarnation of the same was up and running in the form of the Three Little Pigs project. Inspired by

Figure 106. Durakit's "Instant House"—pictured undergoing testing—takes only four days to assemble. Courtesy of F. M. Bartlett.

the nineteenth-century windy fable, the unlikely experimental purpose was to construct a full-scale house to code, then blow it down with full-scale simulated hurricane-force winds. Executing the experimental procedure would be a little more complicated.

The inaugural experiment transpired at a purpose-built $6.8 million facility, located at the periphery of the tarmac at London's airport, about a 20-minute drive from the wind tunnel lab. There the Three Little Pigs house—a two-story, 1,900-square-foot, red-brick house built to code by students from the local community college—sits anchored to a three-foot-deep cement floor and enclosed by a custom-made rolling hangar. The hangar boasts a gigantic bifolding door and rests on tracks that allow this garage of sorts to be wheeled back with the help of a winch, exposing the house to the elements. That the house could be exposed to real weather was only in the interest of some auxiliary tests studying the effects of snow, rain, and mold growth, because as far as wind went, the technological innovations deployed in the Three Little Pigs experiment depended neither on Mother Nature nor on massive Hollywood special-effects fans. Rather, the wind—"the wolf"—was replicated by an orchestra of 100 pressure boxes bolted to the house and attached to a galvanized steel cocoon, a muscular scaffolding that worked as a brace or reaction frame. The pressure boxes—with varying pressures proportionate to box dimensions, two-by-two feet, four-by-four, and eight-by-eight—

Figure 107A. For the Three Little Pigs experiment, investigators constructed a full-scale house and blew it down (or caused failure, at least) with simulated full-scale hurricane-force winds. Courtesy of the Insurance Research Lab for Better Homes.

Figure 107B. The Three Little Pigs team, standing on the roof of the test house. From left, Greg Kopp, David Henderson, Murray Morrison, and Eri Gavanski. Some of the pressure loading actuators that generated hurricane-force wind pressures are visible at top right. The steel beams around the house are part of the structural reaction frame that held in place the airbags through which the wind pressures were applied. Courtesy of the Insurance Research Lab for Better Homes.

were equipped with pressure load actuators, essentially very sophisticated vacuum cleaners (dearer than even a Dyson, the vacuum cleaners consumed half the project's budget to develop and manufacture). Activated in unison, the actuators blasted out pulsating pressures that mimicked the airflow geometry of turbulent gusts as the wind darted around and tore apart a house during a hurricane, and in doing so the actuators also blasted out a drone of 130 decibels—above the pain threshold, but nonetheless investigators could listen to the house cracking apart despite their hearing protection earmuffs.

"When wind blows over a structure, it creates pressures that act on the surfaces of the building," explains Greg Kopp, a Canada research chair in wind engineering and a director at the lab. "As the propeller or jet engine pushes an airplane along, the wing is designed to have low pressures on the upper surface. The pressure difference between the upper surface of the wing and the lower surface causes it to lift into the air. The same thing happens on a house. As the wind moves over it, low pressures are created which cause it to lift," he says. These pressures, being less than atmospheric pressure, are called suctions—the same force that allows a vacuum cleaner to work. "So, instead of blowing wind at the house," says Kopp, "we replicate the pressures and suctions. This is more efficient in terms of power usage, and less expensive."

Before the Three Little Pigs experiment went live, investigators gathered some baseline data, measuring the impact of a hurricane on a miniature house in the confines of one of the lab's boundary layer wind tunnels. This process translated general data on hurricane wind speed into specific data on how those winds excite the geometry of a certain structure. The model was wired with ports to record the pressure on every corner, window, and brick, producing a time history of the turbulent wind pressure accurate to the second. The output from the wind tunnel tests then became the input at the full-scale Three Little Pigs facility—the data programmed into the computer controlled the mechanical vacuums that replayed the pressures and replicated even trace fluctuations in the wind, as many as seven fluctuations per second. Finally, the entire proceedings were captured by about twenty cameras within the house, while thirty-five load cells beneath measured the force of the pressures at play and thirty displacement transducers measured how much the roof lifted off the walls.

When these pseudo-hurricane-force winds bore down, loose nails leapt from their holes and the roof's plywood sheathing peeled off. If the nails were solid, the wind found the next weakest link, perhaps between the trusses and the walls, in which case the whole roof structure lifted and failed. After the roof was torn off, it was reattached with hurricane ties and pummeled again, cranking up the loads and forcing failure to recur. The results from this second round generated leverage with building code committees. "Code committees don't like to change things unless there is an obvious reason," says Bartlett. "Whenever you change a line in the code, then every single designer or engineer who uses that code has to learn something new." The associated cost is great, and thus so is the reluctance to change. "There has to be clear and unambiguous evidence indicating change is needed. It's not just that we can simulate a Category 5 storm," says Bartlett. "We can simulate a Category 5 and see the unsatisfactory performance, and then make changes we think will improve the performance and subject the specimen to exactly the same load. That becomes the evidence that we take to building code committees. We can say: 'This is the way it has always been done, and this is what happens with improvements. This is the change we are proposing, and this is how the performance improves if we make this change.'"

With its preponderance of evidence, the Three Little Pigs project persistently and continually makes the case for reducing risks. From the initial goal of blowing a house down, the research program broadened to include the more all-encompassing goal of understanding every aspect of residential construction. A second test house focuses on precisely how pressures are transmitted down through the roof to the walls, and these investigations are executed in conjunction with a traditional wind tunnel study of a few dozen house shapes—one-, two-, and three-story, and all variety of roof slopes—to merge the best aerodynamic data with the full-scale modeling. Investigators are also alive to the devil in the details, looking at how wall systems' siding and foam insulation stand up in windy regions, which is of particular interest in the United States as energy concerns motivate greater interest in insulating. Kopp recently got the go-ahead for a study on how rain enters through windows, using the same pressure loading actuators but doused with water. And they are conducting studies on how a single piece of roof sheathing, a single piece of plywood, lifts loose. "It all comes down to nails, different nail patterns, different nail types," Kopp says, who notes that much of the damage from Hurricane Andrew, the worst natural disaster in U.S. history, was a result of too many nails on too many houses missing their mark. "Nails are usually spaced at six-inch centers into the edge studs and 12-inch centers for interior studs. By deliberately causing missed nails at different points in the test house, we're looking at how different spacing errors affect the capacity of the sheathing. If you look carefully, you will see that these little, mostly nonengineered products called nails are really important to performance of housing in storms. Engineered nails, like the ring-shank nails, can really improve the performance a lot."

Modeling every and all variations on risk, the evidence feeds in two directions: to the loss-oriented insurance industry—if type A nail is used, then time X is when the roof comes off—and to the cost-saving construction industry, showing that type B nail is better and improves the longevity of the structure. But then, as Kopp notes, "Canadian builders don't like using nails. They'd rather put in staples. So can we find a solution with staples? Houses are really mass-produced, they are a manufactured product. But they don't have the same controls that go into the manufacture of your vehicle. It's a quasi-uncontrolled manufacturing process. We're trying to find ways to allow the optimization to occur, but at the same time to increase the robustness of the construction."

The trick is striking a balance between better modeling of failures for losses for the insurance companies, the amount of acceptable risk set out in the building codes, and what consumers are willing to pay for a solidly constructed house. At present, most North American houses could not withstand the 96-mile-per-hour winds of a Category 1 storm. In fact, the most common source of damaging winds in continental North America, accounting for more than $1 billion in property damage annually, is the simple thunderstorm. And nature isn't so kind as to forfend worse. All it takes is a volatile ascent of warm, low-pressure air to generate a supercell thunderstorm capable of spawning tornadoes.

Tornadoes were another extreme weather entity that Davenport itemized on his International Decade wish list, part of his vision for a Centre for Engineering Investigation of Natural Disasters. And with the Institute for Catastrophic Loss Reduction's support, a pod-sized version recently sprang into action, conducting "forensic engineering" in the wake of significant storms. When a funnel cloud touched down in June 2008 in Tupperville, Ontario, about 16 miles northwest of London, a team ventured out, much as Davenport had in earlier years, to conduct a damage survey. Led by Kopp, the tornado team consisted of three structural engineers and four engineering students, in partnership with Environment Canada meteorologists— "Environment Canada" being the secret password to gaining access to weather crime scenes.

Deployment occurs when Environment Canada gets a report of property damage from a tornado or an eyewitness account. If it's a scientifically worthy case, a call goes out to the lab's dedicated storm-damage-survey cell phone, which one member of the team keeps on his or her person at all times during tornado season. Word then goes out to the other investigators, and they dispatch as soon as possible in what they've dubbed the "Pig Mobile"—a Dodge Caravan equipped with several GPS units, high-resolution digital cameras, and laptop computers with mapping software, as well as tape measures, scales, hard hats, gloves, flashlights, a first-aid kit, and bug spray.

Time is truly of the essence, since well-intentioned though ill-advised storm cleanup often begins immediately—the Good Samaritan neighbor eager to lend a hand with a chainsaw. But just as everything at a crime scene is potentially evidence, all tornado damage is tantamount to scientific data. The primary goal of damage surveys is to ascertain the strength of the winds by virtue of the wreckage. Most of what is known about tornado sizes, their wind speeds and track lengths, is gleaned from this damage left behind, the uprooted trees

Figure 108. Knowledge about tornado wind speeds is mostly gleaned from the damage left behind—for instance, how far a piece of rooftop flew from its point of origin. Courtesy of G. A. Kopp.

and the debris. The forensic engineers conduct autopsies on the wreckage, with an approach a lot like CSI forensics, accumulating small but telling details from the debris field—how far a piece of roof flew from a house, for example. By observing the wreckage, investigators deduce information about the tornado. And by extension, they draw conclusions about the structural performance of buildings and infrastructure, noting winds the structures withstood and winds they could not. At Tupperville, Kopp and his team focused on a medium-sized, wood-framed, metal-clad building that had suffered "global roof failure" even though the tornado's mean winds were below the structure's design wind speed.

While these forensic investigations with tornadoes stemmed from Kopp's desire to generate data to feed to the Three Little Pigs project, the project soon went further afield and included hurricane damage surveys as well. The team headed to New Orleans immediately after Katrina and deploys regularly along the U.S. Gulf Coast, together with another team from the University of Florida. With all the data, investigators connect damage observations in the field with damage observations in the lab, recreating the flight of debris and reenacting failures, both in the traditional wind tunnel and at the Three Little Pigs site. "In the field we see the failures, but we don't see the sequence," notes Kopp. "In the lab we can see the failures right from the very beginning."

In a nutshell, the process of refining knowledge in a back-and-forth between field studies and lab experiments proceeds as follows. Damage surveys reveal problems with structures. Once the investigators have duplicated the debris flight data in carefully controlled wind tunnel experiments, they can correlate debris flight distances with wind speeds. Knowing the wind speeds tells them what the aerodynamic loading was, revealing a building's response to such loads and precisely how failure initiated within the complete structural system. Taking the data one step further, calibrating to the physics of flight and the wind tunnel experiments, investigators develop computer models simulating these debris fields and giving an estimation of wind speed for a much larger range of observations than is cost-effective to actually test. Computer models are also generated to predict structural response to wind load and to develop products and strategies for mitigating the weaknesses and vulnerabilities—products and strategies that can subsequently be tested in the Three Little Pigs facility and then put into practice in construction. Finally, the process comes full circle, back to the damage surveys, to ensure that the mitigation strategies, as implemented in the building codes, ultimately result in the problems being fixed. The cycle takes years, even decades. After Hurricane Andrew in 1992, Florida revised its building codes such that rocks could no longer be used as ballast on rooftops, since they become lethal wind-borne missiles. Only fifteen years hence did damage surveys confirm that these revisions successfully made structures safer.

With the Three Little Pigs project, the lab proved that full-scale hurricane force winds could be replicated with a supertunnel, of sorts.

Until very recently, however, investigators could only wish for technology to accurately replicate tornadoes and downbursts, blasts that spread downward and outward in forceful sheared winds.

Here the lab's work dates to the late 1990s, when the lab proposed a research project with Manitoba Hydro, investigating a significant failure in a major transmission line. Davenport hypothesized that high-intensity localized storm events—that is, downbursts and tornadoes produced by thunderstorm systems—might be responsible for the damage, because the unusual loading produced by downburst winds had the potential for failure at lower wind speeds than for conventional boundary layer winds. Since the lab lacked the facilities to reproduce these storm winds, investigations proceeded numerically, using computational fluid dynamics to reproduce the flow field, applying it to transmission line models, and proving Davenport's thesis correct. "During this research, we also learned that local, intense storm systems, such as tornadoes and downbursts, produce more than 65 percent of damage to buildings and structures in the continental interior North America," recalls Horia Hangan, an engineering professor and a director at the lab. The rash of tornadoes across the United States in 2011 caused over $20 billion in damages. "However, we do not design or test buildings or structures to this type of local high-intensity winds. This is because none of the existing wind tunnels in the world can simulate time-dependent and spatially complex storm events."

"We understood that we have a problem, a huge gap to fill in the research," he says. "We needed to find the tools to do exactly what we do for straight winds in wind tunnels but for high-intensity storm winds. We needed a new type of experiment." Over the course of the Manitoba Hydro project, Hangan made some progress on this front, working in a back room at the lab, cobbling together a tornado and downburst simulator about the size of a shoebox. No model would fit in such a miniature simulator. Hangan was simply trying to recreate some semblance of the high-intensity wind fields. With thunderstorm wind systems, the maximum wind velocity occurs very close to the surface of the earth, at 100 to 130 feet. Scaled down to the 1:500 scale of the traditional wind tunnel, even 150 feet becomes barely a few inches. And Hangan's simulators were much smaller.

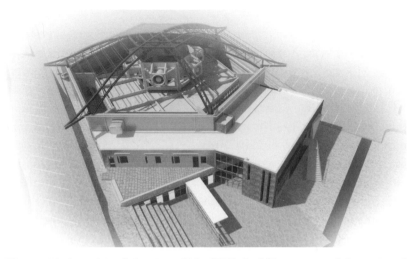

Figure 109. An architect's drawing of WindEEE, the lab's new state-of-the-art Wind Engineering, Energy and Environment facility—the world's first three-dimensional wind tunnel and a bona fide supertunnel. Courtesy of the WindEEE Research Institute.

"Too small," he says. "You couldn't measure anything. We needed something big."

To fit the bill, Hangan conjured up a pie-in-the-sky concept he called WindEEE, for Wind Engineering, Energy and Environment. As the world's first three-dimensional wind tunnel—a bona fide supertunnel—WindEEE is for wind engineers what the Large Hadron Collider is for physicists. It will do what extant wind tunnels cannot.

WindEEE is a dome within a dome, an 82-foot-wide, 26-foot-high hexagonal test chamber housed within a three-storey exterior dome nearly the width of a football field at 131 feet. The test chamber floor plan is divided into concentric zones, and superseding the labor-intensive and crude topographic modeling methods of yore—wood blocks for city and carpeting for water—WindEEE's modular floor system uses a system of digitally controlled roughness elements to generate the desired terrain. The wind tunnel section is studded with more than 100 fans mounted on all six walls as well as the ceiling (where the fans are larger and connected to a plenum box, the opening of which is fast-triggered and mounted on a guillotine quick-traversing system). The surround system fans are orchestrated by sophisticated nonlinear, intelligent control techniques—a particle image velocimetry system measures the velocity of the winds as a feedback loop, correcting errors or discrepancies until the desired conditions are achieved, while the intelligent controls incorporate artificial neural networks. With that Ferarri engine, this sophisticated aeronautical beast has the capacity to generate any type of wind system, such as axisymmetric wind fields, swirling and sheared winds that mimic downbursts and tornados.

Hangan's initial objective with these technological novelties was to invent a versatile tool enabling investigators to ask as yet unanswerable scientific questions about extreme winds, mitigating their destructive powers. The Davenport lab, and labs like it internationally, had mastered the simulation of the turbulent wind behavior in the earth's boundary layer and of large globalized wind patterns, mesoscale systems like hurricanes. But in the nearly fifty years since Davenport's development of the boundary layer wind tunnel, the domain had matured to its limit. And once a domain matures, further progress is limited to technological improvement—in a traditional wind tunnel, for instance, models are now created by rapid prototyping, and their response in the wind tunnel measured by lasers. But with this current technology, localized high-intensity wind systems remained an enigma. "It would be a major breakthrough in wind engineering to achieve these measurements and modeling," says Hangan. "The improvement in public safety and economic benefit would be tremendous."

But, as Hangan likes to note, the wind has two parts, the destructive and the productive. And as WindEEE evolved, its mandate naturally extended beyond the concerns of wind engineering proper to encompass the productive aspects of wind, the ever more urgent issues relating to wind energy, since with climate change and the increase in extreme weather comes an increased demand for alternative and renewable energy sources. The scale and design of WindEEE would also allow investigators to address a host of previously intractable questions about emerging interactions and issues among wind engineering, wind energy, and the environment. For instance, there is the question of why the predicted output for wind energy farms is

Figures 110A–B. To evaluate the performance of this proposed $23.6 million supertunnel with dress rehearsals, project director Horia Hangan constructed a mini WindEEE at a scale of 1:11(above). The viewing window allows investigators to monitor tests in progress, such as flow visualizations of tornadoes using dry ice (below). Courtesy of the WindEEE Research Institute.

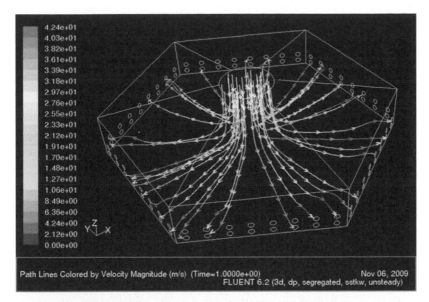

Path Lines Colored by Velocity Magnitude (m/s) (Time=1.0000e+00)   Nov 06, 2009
FLUENT 6.2 (3d, dp, segregated, sstkw, unsteady)

Figures 111A–B. By combining inflow at the ceiling level and radial outflow at the peripheral walls, WindEEE produces downburst-like wind fields (above). When outflow at the ceiling combines with radial and tangential inflow at the peripheral boundaries, the 100+ fans produce tornado-like wind fields (below). Courtesy of the WindEEE Research Institute.

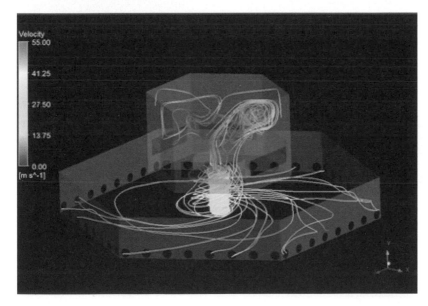

15 percent above actual output, leading to more general questions about the most efficient and effective siting of farms amid topographic terrain. There are questions about the most aerodynamic and durable design for wind turbine blades, about the poorly understood impact of rotational wakes and interference among turbines, and about the effect of energy production on the surrounding environment. These are all questions that could not be addressed within the confines of the traditional wind tunnel but ones that WindEEE will tackle handily.

In composing the Canadian Foundation for Innovation funding proposal, Hangan could not help but use some exultant hyperbole in making his point. This futuristic wind tunnel would be "a highly innovative and versatile facility," "a pioneering study," "an ingenious Canadian invention" that "will truly revolutionize the field of wind engineering." As it turned out, the hype rang true. Construction is nearly completed, and WindEEE is expected to be commissioned and research ready by spring 2013. "It is a once in a lifetime opportunity to build something entirely new," says Hangan. "This is what Alan did with the wind tunnel he founded in the 1960s. It was revolutionary. What we are doing with WindEEE is making the next jump. We are playing and innovating based on the tools that Alan invented. And we hope that WindEEE will be like a toy box for wind engineers for generations to come—to have fun and do something useful at the same time."

Certainly, WindEEE does Davenport proud. He always delighted in thinking outside the box, questing to resolve the unanswerable. And many times during the course of his career he successfully found answers to questions once deemed impenetrable to empirical pursuit. In 2002, at a symposium honoring his life's work, he looked back over his career and borrowed an apropos epigraph from Bob Dylan: "The answer my friend is blowin' in the wind, the answer is blowin' in the wind." There is also another piece of verse, less obvious, that nicely sums up the man and his calling, the steadfast, humble engineer on a humanitarian mission. The stanza, from Matthew Arnold's *Empedocles on Etna*, hung framed just inside Davenport's office door:

Nature, with equal mind,
Sees all her sons at play
Sees man control the wind
And wind sweep man away

It wasn't something to which he drew attention, more something from which he drew inspiration—a wizard's rune, of sorts. Those who did take notice when passing through considered it signature Davenport, a man who simply loved tackling the problems nature presented.

ACKNOWLEDGMENTS

I am indebted to Alan and Sheila Davenport, who generously submitted to endless interviews, follow-up queries, and variegated home invasions, as well as to their children, Tom, Anna, Andrew, and Clare, and to Davenport's brother, Rodney. Many thanks also to everyone from the Alan G. Davenport Wind Engineering Group at the University of Western Ontario who suffered a barrage of inquiries, in particular Peter King, David (and Jean) Surry, Nick Isyumov, and Karen Norman. Paul Kovacs at the Institute for Catastrophic Loss Reduction planted the idea and generously provided funding (as did the lab itself). Les Robertson and SawTeen See educated and inspired on many levels, and opened up the archives at Leslie E. Robertson and Associates (and Karimah Hourmanesh organized the logistics). Ian Pearson gave the final manuscript an insightful once over. Richard Scott read the manuscript at an early stage. Roger Dorton procured the 1992 Canadian Society for Civil Engineering interview with Davenport, answered numerous queries, and shared his photograph of the new Tacoma Narrows Bridge. James Glanz and Eric Lipton of the *New York Times* provided transcripts from interviews with Davenport conducted in 2003 for their book *City in the Sky*. Norman McGrath offered up his official photographs of the Citicorp tower (see www.normanmcgrath.com), and Robert Walker his Amen Corner shots (see www.walkerpix.com). The cover photograph by Ron Nelson came with permission by his son Rob Nelson, also a photographer (www.robnelsonphotography.com). Toronto artist Rick/Simon provided his poster of the CN Tower's "fall radius." At Princeton University Press, thank you to Vickie Kearn, Natalie Baan, and Quinn Fusting. Douglas Bell was ever helpful, especially with syntax and such. Apologies to those I've overlooked. And all that said, all errors are my own.

NOTES

I
Sowing Wind Science

1   "Galloping Gertie": The account of the Tacoma Narrows Bridge disaster and its aftereffects draws primarily from Richard Scott's *In the Wake of Tacoma* (2001) and interviews with Richard Scott.

1   The $6,400,000 bridge: *New York Times* 1940a.

1   "Most features of this disaster": Davenport 1977; Davenport interviews.

4   Dedicated boundary layer wind tunnel: There were, of course, antecedents. In the late nineteenth century to early twentieth century, Denmark's J. O. V. Irminger (about whose work Davenport often raved) devised a tiny wind tunnel driven by chimney exhaust: suction up the chimney drew air through the wind tunnel and over models of houses. Also in the late nineteenth century W. C. Kernot in Melbourne, Australia, designed a more sophisticated wind tunnel driven by a fan, blowing air over models on a balance to measure forces on a wide-ranging scope of designs, including buildings as well as isolated elements such as parapets and overhangs. In the mid-twentieth century, the National Physics Laboratory in Teddington, England, tested primarily airplanes, and to that end mounted some boundary layer tests in their aeronautical wind tunnel. Also during the 1950s Martin Jensen at the Technical University in Denmark built a boundary layer wind tunnel in which he conducted full-scale and model comparisons on one small building, but he mostly focused on shelter belts, such as hedges and trees, investigating the wind protection they provided for crops.

5   "Augusta National is a one-of-a-kind golf course" ("Driving into the Wind" sidebar): The account of the Augusta National tests draws primarily from Shipnuck 2002, as well as interviews with Greg Kopp.

7   Spray fruit trees: Davenport interviews. Davenport conducted this research on behalf of his University of Western Ontario colleague Ion Inculet.

8   A powerful lashing of wind: *Canadian Geographic*, February 1987.

10   Efforts to cut sulfur emissions from chimneys ("Upstream, Down-stream" sidebar): The account of tests on stacks and chimneys draws from interviews with Barry Vickery.

11   "For they have sown": Davenport 1979.

13   The wind was first treated: Ibid.

14   Beaufort Wind Force Scale: Huler 2004.

14   Beautiful solutions: Davenport interviews; David Surry interviews.

15   Concept of a boundary layer: Davenport 1977, 1979; Davenport and Surry interviews.

15   "Magical": Surry interviews.

16   "Now to what phenomenon": *Le Temps* 1887.

16   "With regard to the exposed surfaces": Quoted in Davenport 1977.

17   Eiffel's experiments on subjects: Davenport 1977.

18   "Certainly no designer": Henri Petroski interviews.

18   "It is interesting to consider": Davenport 1999b.

18   A brief recap of formative years: Davenport interviews; Sheila Davenport interviews; Rodney Davenport interviews.

23   "It was a puzzle": White 1992 (Davenport interview with the Canadian Society for Civil Engineering, June 1992).

24   "There are two principal differences": Ibid.

27   "Turbulence was an aspect": Ibid.

27   "Suddenly what burst": Ibid.

28   The 237-page document: Davenport 1961b.

28   "Transference was the key": Ibid.

30   A naïve yet crucial guess: Allan McRobie interviews; Sir Julian Hunt interview.

II
Tall and Taller Towers

32   Answered the call: Davenport interviews.

32   "Singing a duet": Glanz 2003, Glanz and Lipton 2003.

32   Davenport could hardly help but swoon: Ibid.

32   "Fell off several chairs": Glanz 2003; Davenport interviews.

32   Meeting at Hotel London: Davenport interviews; Leslie Robertson interviews.

33   Father of modern wind engineering: The International Association of Wind Engineering, after "interpreting the words and feelings of several

NOTES • 231

colleagues of every part of the world," in 2007 established the Davenport
Medal as the senior IAWE award. In doing so, IAWE president Giovanni
Solari noted that "other fathers of wind engineering already link their
names with outstanding recognitions. For instance, ASCE periodically
awards a 'Cermak' medal and a 'Scanlan' medal. Following two IAWE
resolutions, we commonly use the names 'Jensen Number' and 'Scruton
Number'. An analogous recognition is not yet present for Prof. Davenport, widely recognised as a symbol of modern wind engineering."

33   Wind engineering giant was awakening: Davenport interviews.
33   Two papers: Davenport 1965a, 1965b; Davenport interviews.
33   Questioning by someone from the Met Office: Davenport interviews.
34   Jerry-rigged experiment: Davenport 1965a, 1965b; Davenport inter-
     views.
34   "Sometimes the wind blew": Davenport 1969a.
35   Unlike anything ever built before: Davenport interviews; Robertson in-
     terviews.
36   "Paramount in importance": Quoted in Davenport 2001.
36   "I went seeking genius": Robertson interviews.
37   "The well-meant efforts": Schuyler 1909.
37   A change in technology: Davenport 1977, 2002; Khan 2004.
37   "Form ever follows function": Sullivan 1896.
38   Tube structure: The invention of the tube structure is also attributed to
     Chicago engineer Fazlur Khan, who concurrently invented a similar con-
     cept first applied to the Dewitt-Chestnut Apartments; see Khan 2004.
38   Wind speed concepts: Davenport interviews, Nicholas Isyumov inter-
     views, Peter King interviews; Surry interviews.
39   Weren't the least bit credible: Davenport interviews.
41   Davenport's unprecedented wind engineering innovations: Davenport
     interviews; Surry interviews.
41   "The engineering art": Davenport interviews.
42   "The Model Law": Jensen 1958; Davenport interviews; Surry interviews.
42   Cermak was curious about Davenport's intentions: The account of the
     World Trade Center (including a composite account of wind tunnel tests,
     combining details from the CSU and UWO studies) draws from Daven-
     port 1969a, 1969b, 1975, 2001; Davenport, Isyumov, Fader, and Bowen
     1970; *Engineering News-Record* 1980; Glanz 2003; Glanz and Lipton
     2003; Miller and Murphy 1980; Seabrook 2001; Worthington, Skilling,
     Helle & Jackson 1966; and from interviews with Davenport, Isyumov,
     Robertson, and Surry.

44 "Wind engineering is the rational treatment" (caption, figure 25): Cermak 1975.

49 "Twentieth-century Gullivers": quoted in Glanz and Lipton 2003.

50 "Huge amplitudes": White 1992.

50 "Pick a number": Ibid.

51 "We knew nothing": Ibid.

51 Davenport's quest to quantify the wind: Davenport interviews; Robertson interviews; Surry interviews.

52 "It's not exactly Christopher Columbus": Robertson interviews.

52 "If we added turbulence": White 1992.

53 Hit the dance floor: Robertson interviews.

54 Robertson's unprecedented precautionary step: Chen and Robertson 1972; Glanz and Lipton 2003; Davenport interviews; Robertson interviews.

54 Promised complimentary checkups: Glanz and Lipton 2003; Robertson interviews.

55 "I am going to project": Quoted in Glanz and Lipton 2003.

55 Six comfort zones: Chen and Robertson 1972.

56 "Whoa boy!": Quoted in Glanz and Lipton 2003.

56 "Reeling around": Ibid.

56 "I'm on a boat": Ibid.

56 "Taking away my gravity": Ibid.

56 "It's unpleasant": Ibid.

56 Over two weeks: Chen and Robertson 1972; Glanz and Lipton 2003; Robertson interviews.

56 "Rounded feet": Quoted in Glanz and Lipton 2003.

56 Wind tunnel tests continued: Davenport interviews; Isyumov interviews; Surry interviews.

57 "I was very insistent": White 1992.

57 Juggling a bevy of variables: Davenport interviews; Isyumov interviews; Robertson interviews; Surry interviews.

58 "Important consequences in the design of the towers": Quoted in Glanz and Lipton 2003.

58 Optimized structures: Robertson interviews.

58 "It is evident that the response of the towers": Quoted in Worthington, Skilling, Helle & Jackson 1966.

59 "Davenport's Probabilistic Prediction Method" (sidebar): Surry interviews.

59 "Because of the directional characteristics of wind" ("Davenport's

Probabilistic Prediction Method," sidebar): Quoted in Worthington, Skilling, Helle & Jackson 1966.

64   Tweaking of the towers' performance: Davenport interviews; Robertson interviews.

64   "We told him in Detroit": Robertson interviews.

64   "We wouldn't have to worry": White 1992.

64   Friction dampers under the floor: Davenport interviews; Robertson interviews.

64   "The apparatus was outside of their ken": Robertson interviews.

66   Wonder whether the ultimate design: Robertson interviews.

66   But the dampers did more than buffer the effects of wind: Glanz and Lipton 2003; Davenport interviews; Robertson interviews.

66   Robertson broke down and wept at the lectern: Seabrook 2001.

69   Clean conceptual equation: Davenport interviews; Isyumov interviews; Surry interviews.

69   "Davenport's Wind Loading Chain" (sidebar): Isyumov 2011; Isyumov interviews.

71   Jet-setting had its hassles: Davenport interviews.

73   Davenport's laboratory: Initially called the University of Western Ontario Boundary Layer Wind Tunnel Laboratory, the lab was officially renamed the Alan G. Davenport Wind Engineering Group on Davenport's retirement in 2002.

73   Ferrybridge Power Station cooling towers collapse: Davenport interviews; King interviews; Isyumov interviews; Surry interviews.

77   The triangle with indented corners: Davenport 1969b.

78   "Skyscraper fever": *Engineering News-Record* 1966.

78   Business was booming: The account of the Sears project draws from Davenport, Isyumov and Jandali 1971; Khan 2004; Miller and Murphy 1980; Davenport interviews; Isyumov interviews; Surry interviews; John Zils interview.

82   Hundredth of an inch: The Sears Tower model was constructed by skilled artisans recruited from Europe, under the direction of Dutchman Bill Ramaekers, who not long before had established the machine shop at the University of Western Ontario's Faculty of Engineering.

82   "It looked like a Rube Goldberg machine": Zils interview.

86   At 1,815 feet: The account of the CN Tower project draws from Isyumov and Davenport 1975; Isyumov, Davenport, and Monbaliu 1984; Isyumov et al. 2000; Davenport interviews; Isyumov interviews; Knoll interview.

91   The long-span tensile membrane roof of the Calgary Saddledome: The account of the Saddledome project draws from Davenport and Surry 1984; Davenport interviews; Surry interviews.

92   The fabric roof designed for the Haji Terminal: The account of the Hajj project draws from Tryggvason, Surry, and Davenport 1979; and interviews with Davenport, Surry, and Bjarni Tryggvason.

98   "One Tower, Two Tuned Mass Dampers" (sidebar): The account of the Sydney Tower project draws from interviews with Vickery.

101   "The Holy Grail of Full-Scale Testing" (sidebar): The account of the Allied Bank project draws from Isyumov and Halvorson 1984; and interviews with Davenport, Isyumov, Surry, and Robert Halvorson.

107   The Council on Tall Buildings and Urban Habitat: The council is now located at the Illinois Institute of Technology in Chicago.

107   "The dominant influence": Miller and Murphy 1980.

107   "A bonanza for a handful of experts": *Engineering News-Record* 1980.

107   "None of these studies was routine": Davenport 1969a.

107   On a stormy night in January: The account of the John Hancock, Boston, project draws from Campbell 1995; Davenport, Surry, and Tanaka 1975; Surry, Tanaka, Allen, and Davenport 1975; Thürlimann 1975; and interviews with Davenport, Isyumov, Surry, and Hiroshi Tanaka.

108   "Like sequins off a dress": Campbell 1995.

112   "Davenporting": Surry interviews; King interviews; Vickery interviews.

113   The next calamity: The account of the Citicorp project draws from Davenport 1996; *Engineering News-Record* 1995; Morgenstern 1995; and interviews with Davenport, Isyumov, Surry, and Robertson.

114   "Tottering tower": Morgenstern 1995.

116   "The Wind, Fickle and Shifty, Tests Builders": Horsley 1974.

116   "A classroom lecture": Quoted in Morgenstern 1995.

117   "If we are going to think about such things as the possibility of failure": Ibid.

118   "I had information that nobody else had": Ibid.

120   Codes also steered a conservative course: The account of the Hong-kong and Shanghai Bank project draws from Davenport, Isyumov, and Bowen 1970; Davenport, Isyumov, and Greig 1978; Davenport, Surry, and Lythe 1984; Lythe and Surry 1982; Tschanz and Davenport 1983; Williams 1989; and interviews with Davenport, Robertson, and Surry.

121   "The brief for the Hongkong and Shanghai Bank Headquarters": Quoted in Williams 1989.

125   Shanghai World Financial Center: The account of the Shanghai World Financial Center draws from interviews with Davenport, Isyumov, and Robertson.

126   Even the Burj Khalifa: The account of the Burj Khalifa draws from interviews with Peter Irwin and Davenport.

III
Long and Longer Bridges

129   Bridges were his first structural love: Accounts of Davenport's philo- sophical and catholic interests draw from interviews with Sheila Daven- port and the Davenport children, Tom, Anna, Andrew, and Clare.

131   Calatrava, the architect who doubles: Pollalis 1999; and King inter- views.

133   "Architecturally the finest suspension bridge of them all": *New York Times* 1939.

133   "Breaks no records for length of span": Quoted in Barelli, White, and Billington 2006.

133   Deflection theory of suspension bridges: Davenport interviews; King interviews.

134   Roebling recipe: Petroski interviews.

134   "What drove the changes": Ibid.

134   "They were narrow": Ibid.

135   From the ruin: Scott 2001.

136   "Like lifting a man up by his suspenders": Moses 1970.

136   Warren trusses: Barelli, White, and Billington 2006; Rastorfer 2000.

136   "The britch is safe": Quoted in Moses 1970.

136   "Extraordinarily small width of structure": Quoted in Scott 2001.

138   "Their thinking was still largely influenced by consideration of static forces": Ibid.

138   "[The report] leaves many questions unanswered": Ibid.

138   The aerodynamic instability of bridges would become a going con- cern: Scott 2001; Davenport interviews; King interviews; Surry inter- views. Following the Tacoma report, another commission was struck to continue investigations, with an eye to the design of the replacement, and to the aerodynamic design of suspension bridges generally. But these in- vestigations, too, left the phenomenon that felled the bridge lacking co- herent explanation. It was Robert H. Scanlan at Princeton University

who provided the first plausible explanation. After working on the problem for the better part of forty years, he argued convincingly that the failure was due to flutter. Scanlan, who became known as "the father of flutter," did not have a wind tunnel at his disposal and worked almost exclusively by theory, in stark contrast to Davenport, a devout experimentalist as well as a prolific theoretician. Scanlan's work provided what is now a standard analysis for long-span bridges, though wind scientists would continue to debate competing theories regarding the Tacoma's failure for years to come. The latest study, published in 2000 by the Danish bridge aerodynamicist Allan Larsen, concluded that vortices emanating from the leading edge of the H-girder also played a crucial role in the bridge's final moments.

139 Preceded by his reputation: Roger Dorton interviews.

139 "Buffeting of a Suspension Bridge by Storm Winds": Davenport 1962a; Davenport interviews.

140 "Participants journeyed north": National Physical Laboratory 1965.

140 Keen verging on reckless enthusiasm: Davenport interviews.

140 "Are you going to do any wind tunnel testing?": Davenport interviews.

142 "Components of Davenport's Wind Theory" (sidebar): Davenport interviews; King Interviews; Surry interviews.

146 A. Murray MacKay Bridge: The account of the wind tests on this project draws primarily from Davenport and King 2000; King 2003; and interviews with Davenport, Dorton, and King.

148 "Davenport saved the day": Dorton interviews.

150 "All suspension bridges sway": Quoted in Perlmutter 1968.

150 "Bridge doctor": Finn 2000.

150 Tests on the Tacoma Narrows replacement bridge: Scott 2001. In fact, using an ingenious but crude array of electromagnets, Farquharson had been in the midst of tests on a 54-foot-long model of the original Tacoma Narrows Bridge when it fell.

150 Bronx-Whitestone Bridge: The account of the wind tests on this project draws primarily from Barelli, White, and Billington 2006; Davenport and Isyumov 1982; Davenport, Isyumov, and Tanaka 1976; King, Kopp, Kong, et al. 2000; King, Morrish, Isyumov, et al. 2000; Rothman 1995; Velez and Fanjiang 2005; and interviews with Davenport, Isyumov, King, Surry, and Rothman.

151 "This is typical of a bridge made torsionally unstable": Rothman 1995.

151 Not wanting the U.S. federal government to get wind: Rothman interviews.

154 Davenport's taut strip model: Accounts of the taut strip model draw

primarily from Davenport 1972; Davenport, King, and Larose 1992; and
interviews with Davenport, King, Isyumov, Surry, and Rothman.

158   "It had been such a mystery": Rothman interviews.

158   "A slight probability": Rothman 1995; Rothman interviews.

158   Propeller-toaster damper: Rothman interviews.

159   Damper was more complex: King interviews; Rothman interviews.

159   Sunshine Skyway Bridge: The account of the wind tests on this project
draws primarily from Davenport and King 1982a, 1982b; Davenport,
King, and Larose 1992; Sayers 2008; and interviews with Davenport,
King, and Surry.

161   "Up until then, the only component": King interviews.

162   "Take a steel ruler": Ibid.

165   "Most bridge engineers want one load": Ibid.

167   "We can see the proof": Figg interviews.

168   Tipped off by acoustic sensors: King interviews; Velez interviews.

168   As Hurricane Floyd blew into New York: King interviews.

170   The accrual of full-scale data: Davenport 1999b; Davenport and King
interviews.

171   "Then one doesn't have to wait": King interviews.

172   "We put this bridge on a diet": Quoted in Chan 2005.

172   "The answer was not good": Velez interviews.

172   Configurations for fairings: King interviews.

173   Redecking the bridge: Ibid.

175   Whitestone transformation complete: Ibid.

175   "I'd tell you it should last forever": Velez interviews.

175   Pilgrimage from bridge to bridge: Davenport interviews; Sheila Davenport interviews; King interviews; Surry interviews.

177   "We climbed out at a dizzying height": Sheila Davenport 1998.

177   The lore about the superstructure's finer details: Miyata and Yasuda
1993; Kashima and Kitigawa 1997.

178   Miyata proudly described the end result of his labors: Miyata and Yasuda 1993.

179   "Snake-like motions": Ibid.

179   Strait of Messina Bridge: The account of this project draws primarily
from interviews with Davenport, King, and Larose. See also the Aconex
project case study web page, http://www.aconex.com/case-studies/strait-
messina-bridge.

182   "A Bridge Too Far": *Economist* 2003.

IV
Project Storm Shelter

183 Weather warning bulletin: Ontario Climate Centre 1996.

184 Davenport had never been a storm chaser: Davenport interviews.

185 "It's almost an epidemiological approach": Ibid.

185 International Decade: Davenport interviews; Frank Press interview.

186 "This special initiative would see all nations": Quoted in *International Association for Earthquake Engineering 1992.*

186 UN Resolution 236 announced: United Nations 1989.

187 The mandate was clear: Canadian National Committee for the International Decade for Natural Disaster Reduction 1994; Royal Society of Canada 1993; United Nations 1989.

187 Natural disasters had killed three million people: Royal Society of Canada 1993.

187 "History and mythology are filled with tales": Canadian National Committee for the International Decade for Natural Disaster Reduction 1994.

187 Seismic zoning map: Milne and Davenport 1969. Davenport further pursued this interest, first in the 1980s with PhD student Gail Atkinson and the lab's authority on foundational dynamics Milos Novak, analyzing different methods for calculating seismic probability and applying the analysis to determine the probability of damage to infrastructure such as the TransCanada pipeline, and most recently publishing one of his last papers, in 2006, co-authored with PhD student Katsu Goda and H. P. Hong of Western's Civil and Environmental Engineering Department, a comparative study of seismic hazard analysis by the Davenport-Milne method, the Cornell-McGuire method, and the epicentral cell method, and how these methods work together to give the most nuanced picture of hazard.

188 "Some of the most populated areas in the world": Davenport, Georgiou, and Surry 1986.

189 Hurricane wind risk studies: Tryggvason, Davenport, and Surry 1976.

190 Predicting cyclone wind speeds using the Monte Carlo method: Davenport, Tryggvason, and Surry 1976.

191 "In essence this so-called 'Monte Carlo'": Georgiou, Davenport, and Vickery 1983.

191 "The historical data records": Davenport, Georgiou, and Surry 1986.

192 Hurricane risk map for the entire Caribbean: Davenport 1992a; Davenport, Georgiou, and Surry 1986; Gibbs 2002.

192 Following Hurricane Gilbert: Gibbs 2002. Davenport presented his findings in 1989 at a seminar titled "A One-Day Course on Wind Loads on Buildings—Caribbean Uniform Building Code," held in Jamaica and cosponsored by the Jamaica Institution of Engineers and the Canadian Society for Civil Engineering.

193 "Eschewed dogmatism": Gibbs 2002.

195 $32 billion in damages: Canadian National Committee for the International Decade for Natural Disaster Reduction 1994.

195 Mapping the vulnerabilities: *National Geographic* 1998.

195 "Center for engineering investigation": Canadian National Committee for the International Decade for Natural Disaster Reduction 1994; Davenport interviews.

195 "Double tragedy following natural disasters": Davenport 2000c.

198 "The United Nations is observing the day": National Public Radio 1995.

198 "We confront a paradox": United Nations 1999.

199 "Do not build": Quoted in Fackler 2011.

199 Elephant matriarchs draw on memories of hard times: Revkin 2008.

199 Low-probability problems: Davenport interviews.

199 "If you do it right, nothing happens": Gordon McBean interviews.

200 UN International Strategy for Disaster Reduction: Ibid.

201 "Ingrained in Alan was the belief": Paul Kovacs interviews.

201 Institute for Catastrophic Loss Reduction: Davenport interviews; Kovacs interviews.

201 Low buildings typically struggle to compete: Davenport and Surry 1974; Davenport, Surry, Stathopoulos, et al. 1978; Davenport interviews; Surry interviews.

201 One of the key issues: Davenport 1999b.

202 "Dangerous and dirty": Ibid.

202 Metal Building Manufacturers Association study: Davenport interviews; Surry interviews.

203 "If you just measure at one point": Surry interviews.

204 "Averaging the Air" (sidebar): The account of the invention of the pneumatic averager draws from interviews with Ted Stathopoulos and Surry.

206 "Raising the Roof on Low Building Testing" (sidebar): The account of the innovative techniques for low buildings draws from interviews with Stathopoulos and Surry.

207 "Guiding principle Davenport called 'total quality management'": Davenport 1999b.

208 "It buffers the risk of all the stakeholders": Ibid.

208 Project Storm Shelter: Davenport 2000c.

208 "Catalogue of pressure coefficients": Davenport 1999b; Davenport interviews; Surry interviews. The best coverage of shapes at the time entailed insightful yet outmoded studies by Martin Jensen and Niels Franck, and subsequently various but more limited studies by D. Surry, T. Stathopoulos, E. Ho, and G. M. Richardson at the Davenport lab, as well as work by John Holmes in Australia.

209 Supertunnel: Davenport 1985, 1999b; Davenport interviews. Davenport first presented the idea of a supertunnel at a conference in Roorkee, India, in 1985. The National Academy of Engineering in the United States was considering a similar concept—the "Wall of Wind" hurricane machine—and established a blue ribbon commission, on which sat Davenport and Jack Cermak, among others, to investigate the possibilities. At one juncture, NASA was considered a potential partner, allowing for the Ames aircraft wind tunnel to be used for this purpose. And subsequently, Sir Julian Hunt, a theoretical wind engineer in the United Kingdom who later became the Astronomer Royal, arranged for a conference, "Dealing with Natural Disasters," in 1999 at the Royal Society in London, where Davenport spoke on progress during the International Decade, as well as his "thought experiment" regarding a supertunnel. But alas, his supertunnel in that era proved financially out of reach.

209 "Thought experiment": Davenport 1985, 1999b.

210 "Damage quotient": Davenport 1999b; Davenport interviews; Surry interviews.

210 Make its name with a big-impact project: Bartlett interviews; Kopp interviews; Surry interviews.

210 Durakit: D'Costa and Bartlett 2000, 2003; Bartlett interviews.

210 Three Little Pigs project: Kopp et al. 2010; Roberts 2009; Kopp interviews. The prototype for the Three Little Pigs was developed in the late 1980s by the Building Research Establishment in the UK. The BRE (equivalent to Canada's National Research Council) had been a hub of study on low-building problems (sometimes by instrumenting full-scale structures) dating back to early days. RWDI-Anemos director Nicholas Cook, a Bristol alum, like Davenport, worked at the BRE's modern lab and developed with Cambridge Consultants, an engineering firm that prides itself on creating "the killer product," the BRErWulf device—"BRE Realtime Wind Uniform Load Follower"—which was the model for the pressure box devices that the Davenport lab deployed for the Three Little Pigs project.

213 "When wind blows over a structure": Kopp interviews.

214 "Code committees don't like to change": Bartlett interviews.

214 Alive to the devil in the details: Kopp interviews.

215 "It all comes down to nails": Ibid.

216 Tornadoes were another extreme weather entity: Ibid.

218 "In the field, we see the failures": Ibid.

219 Significant failure in a major transmission line: Horia Hangan interviews.

219 $20 billion in damages: Seelye 2011.

219 "We do not design or test buildings": Hangan interviews.

220 "We needed something big": Ibid.

220 WindEEE: The account of the WindEEE project draws from Hangan 2008; Hangan interviews.

224 "A highly innovative and versatile facility": Hangan 2008.

224 "It is a once in a lifetime opportunity": Ibid.

225 "Nature, with equal mind": The Arnold poem was given to Davenport by an early student, Rein Lemberg, now vice president of risk management at CALTROP.

INTERVIEW SOURCES

Over the course of researching and writing *Wind Wizard*, from 2007 to 2011, I conducted numerous interviews with primary and secondary sources via phone, email, and in person. Alan Davenport submitted to several rounds of interviews prior to his death on July 19, 2009. Sheila Davenport participated in several interviews and answered endless follow-up questions. The Davenport children, Tom, Anna, Andrew, and Clare, were generous with their reminiscences, as was Alan Davenport's brother, Rodney Davenport. At the Davenport Wind Engineering Group, Peter King and Dave Surry generously vetted many drafts of the manuscript and were crucial sources, as were Nick Isyumov, Barry Vickery, Greg Kopp, Horia Hangan, Mike Bartlett, and Karen Norman.

Additional interviews for Part I included Hentry Petroski, Duke University; Richard Scott, National Capital Commission; Allan McRobie, Cambridge University; Sir Julian Hunt, University College London; and Gordon H. Smith, Gordon H. Smith Corp.

For Part II, additional interviews included Les Robertson and SawTeen See of Leslie E. Roberston Associates; John Zils, Skidmore, Owings & Merrill; Bjarni Tryggvason, University of Western Ontario; Franz Knoll, Nicolet Chartrand Knoll; Hiroshi Tanaka, University of Ottawa; Peter Irwin, Rowan Williams Davies and Irwin Inc.; Mike Hogan, Hogan & Greenfield Design-Build Ltd; Alan Dalgliesh, consultant formerly of National Research Council of Canada; Robert Halvorson, Halvorson and Partners.

For Part III, additional interviews included David Billington, Princeton University; Richard Scott, National Capital Commission; Roger Dorton, retired engineer; Guy Larose, National Research Council Institute for Aerospace Research; Toshio Miyata, Yokohama National University; Linda Figg, Figg Engineering Group; Herb Rothman, formerly of Weidlinger Associates; Genaro Velez, Weidlinger Associates.

For Part IV, additional interviews included Paul Kovacs, Institute for Catastrophic Loss Reduction; Frank Press, Huron Consulting Group; James Bruce, assistant deputy minister, Meteorological Service of Canada, and chair, Scientific and Technical Committee, International Decade for Natural

Disaster Reduction; Gordon McBean, University of Western Ontario, and chair of the Integrated Research on Disaster Risk, International Council for Science and International Social Sciences Council; Gail Atkinson, University of Western Ontario; and Ted Stathopoulos, Concordia University.

Interviews with Davenport's colleagues internationally included Steve Killing, Steve Killing Yacht Design; Brian Smith, Flint & Neill Partnership; Prem Krishna, University of Roorkee; Tony Tschanz, Magnusson Klemencic Associates; and Manabu Ito, the University of Tokyo.

GLOSSARY

Aerodynamic admittance: The efficiency with which a structure converts the energy in the wind to a force acting on the structure, including the effect of the correlation of flow in the vicinity of the structure.

Aeroelasticity: A property resulting from the combination of aerodynamic effects and the elasticity of a structure. For example, the twisting of a structure may lead to a change in the aerodynamic loads, which in turn leads to further twisting of the structure.

Axial deformation: The deformation that occurs along the axis of a member as a result of stresses placed on the member. For example, a column in a tall building supports a great deal of weight, which leads to it being slightly shortened.

Collapse modes: The different possible ways for a structure to collapse. A house may collapse because the roof is lifted off, leaving the walls with little support; or the walls may collapse, allowing the roof to fall in.

Critical velocity: The wind speed associated with the onset of a structure's instability, for example, the velocity at which wind induces flutter instability in a bridge.

Damping: The dissipation of energy in a structure oscillating in one of its modes of vibration. It is normally expressed as a damping ratio.

Damping ratio: The ratio of the damping to the critical damping, which is the least amount of damping required to suppress any oscillatory response.

Degrees of freedom: A property of structures that respond in multiple ways when excited, such as a building responding in the two sway directions and in torsion, and perhaps in multiple modes of vibration in each direction.

Design wind speed: The speed chosen by the designer for a structure's strength and performance calculations. The forces generated at the design wind speeds incorporate safety factors, and as a result, the failure wind speed is considerably higher than the design wind speed.

Displacement transducers: Instruments used to measure displacements or deflections.

Dynamic amplification: An increase in motion that occurs when a vibrating structure is excited at one of its natural frequencies such that a small amount of energy input is in phase with the motion. This can cause the motion to become increasingly large, limited only by the amount of inherent damping in that mode.

First mode vibration: A structure's motion at its lowest natural frequency; the associated shape of vibration for a building can often be approximated by a straight line rotating at the base, while that for a transmission line or a bridge would be a half sine wave. All bodies have preferred frequencies at which they vibrate, called natural frequencies. Each natural frequency corresponds to a vibration essentially having a constant shape, called a mode shape, that grows and decays in amplitude with time.

Flutter: An instability in flexible bodies caused by the interaction of the wind and the structure, involving either pure torsional motion or coupled vertical and torsional motion, and involving oscillations with amplitudes that increase in time and can result in catastrophic structural failure.

Frequency: The number of times per unit time that an event occurs. For example, electrical power in North America is transmitted as AC or alternating current at 60 Hz—sixty cycles per second of voltage and current.

Hurricane ties: Straps designed to connect elements of buildings together to better resist hurricane loads, such as tying roof elements to wall elements. They are thin strips of steel that wrap over the top of the roof truss and are nailed to the wall studs, significantly strengthening the connection and adding at most a few hundred dollars to the cost of a new house.

Influence lines: Conceptual lines that relate some response of a structure to the load at an arbitrary point along the line. For example, with a cantilever beam, the influence line for deflection at the tip would look a lot like its shape when loaded at the tip. That is, loads on a cantilever produce no deflection at the tip when the load is at the base, but produce more and more tip deflection as the load is moved toward the tip. The concept can also be extended to an influence surface, where the load can be at an arbitrary three-dimensional point on the structure.

Joint mode acceptance (or joint acceptance function): A mathematical relationship that couples the forces acting on a structure with the modes of vibration of the structure.

Lateral mode amplitude: Amplitude of a mode of vibration (for example, of a bridge deck) that involves predominantly lateral or transverse motion relative to the longitudinal axis of the bridge.

Lateral mode frequency: A natural frequency of a mode of vibration (for example, of a bridge deck) that involves predominantly lateral motion.

Load cells: Instruments designed to measure loads applied, just as a kitchen scale measures weight. Load cells can be designed to independently measure loads applied in more than one direction at the same time.

Longitudinal undulation: A sinuous vertical movement along the length of the structure.

Lumped-mass rigid-floor diaphragm: A rigid plate used in the design of an aeroelastic model. When an aeroelastic model is designed for a building it is often simplified by dividing it into a number of layers, each layer corresponding to a number of floors in the full-scale building. The mass of each collection of floors is modeled as a simple rigid plate of metal with a correctly modeled mass and mass moment of inertia. These lumped masses are then connected together by thin columns designed to provide the correct model of the building's elasticity.

Mass moment of inertia: A measure of a body's inertial resistance to rotary motion, similar in concept to the moment of inertia. It is the integrated sum of each element of mass of the cross section of the body, multiplied by the distance from the axis of rotation squared.

Mode: The dynamic motion of a structure as it resonates at one of its natural frequencies—that is, as it vibrates back and forth, in positive and negative directions.

Modal frequency: The natural frequency of a particular mode of vibration. All bodies have preferred frequencies at which they vibrate, called natural frequencies, which are dependent on the stiffness and mass distribution of the structure. Each natural frequency corresponds to a vibration essentially having a constant shape, called a mode shape, that grows and decays in amplitude with time.

Mode shape: The shape a structure deflects or deforms into when vibrating at a specific modal frequency.

Moment of inertia: A measure of a body's resistance to bending. It is the integrated sum of each element of area of the cross section of the body, multiplied by the distance from the axis of bending squared.

Natural vibrating frequency: The preferred frequency at which a body vibrates.

Overall structural loading coefficients: A dimensionless coefficient that conveys the effective load on different parts of the structure. This innovation was motivated by the fact that for the first time, the overall loads on large parts of the model buildings were measured directly, capturing the way the unsteady pressures averaged out over large parts of the structure, the principle being that the larger the area, the more averaging out of the unsteady loads would occur, and thus the associated design loads could be reduced.

Pneumatic averaging: A simple technique for spatially averaging the unsteady pressure over an area. Rather than using a single pressure tap to get a representative pressure for an area, multiple identical pressure lines can be used and connected together using a manifold. The result is a physical device for determining the spatially averaged pressure. If done correctly, most of the correct unsteady aspects of the individual component pressures are also averaged properly. Davenport extended this idea by using a porous material over the area of interest to allow the porosity to replace individual pressure lines. The method has now largely been replaced by digital averaging of simultaneously sampled signals.

Pressure coefficient: A nondimensional representation of pressure, generally formed by dividing the pressure by the dynamic pressure of the wind. The latter is 0.5 times the fluid density times the mean wind speed squared, and corresponds to the pressure that is recovered when wind is brought to a halt. Pressure coefficients can be measured in the wind tunnel and transferred directly for full-scale use in most cases.

Random buffeting forces: The loads generated in a structure by turbulence in the natural wind.

Rapid prototyping: A technology used to make a single complicated shape rapidly. For example, before committing to producing millions of telephones with expensive molds, it is useful to make a prototype to check the details. These technologies have been adapted to making building models. One example is a computer-controlled nozzle that lays down a tiny stream of molten plastic in layers only about 1/50th of an inch thick. It builds up the shape in repeated layers, changing the shape of each as required.

Resonant frequency: The frequency at which a system gains energy from an external excitation. For example, when a tower is subjected to dynamic loading that is in phase with any one of the structure's natural frequencies, this will cause the structure to resonate (if it lacks critical damping). Wind turbulence includes fluctuations over a wide range of frequencies such that there is always some energy that will match a particular structural natural frequency, which then is a resonant frequency.

Response or gust factor: A factor that, when multiplied by the mean (or time-averaged) wind force, provides an estimate of the maximum dynamic wind load. The gust factor is a function of the structure's natural frequency.

Self-damping: The inherent damping or energy dissipation that occurs in any structure when caused to vibrate.

Stay: A taut cable, as in a "cable-stayed" bridge. A cable stay is constructed out of a number of solid or twisted wires that are bundled together to act as a single tension element.

Stiffness: The resistance of a structure to bending.

Torsional acceleration: Acceleration in a mode of vibration (for example, of a bridge deck) that involves predominantly torsional motion. This is commonly defined as a vertical acceleration at the edge of the bridge deck, or as an angular acceleration about the longitudinal axis of the bridge deck.

Torsional mode amplitude: Amplitude of a mode of vibration (for example, of a bridge deck) that involves predominantly torsional motion.

Torsional mode frequency: A natural frequency of a mode of vibration that involves predominantly torsional motion.

Torsional motion: Vibration in which the structure twists back and forth relative to the longitudinal axis of the bridge.

Torsional vibration: A vibration in which the structure twists back and forth about an axis of rotation.

Tributary area: The area that contributes load to a member of interest. On a roof, the tributary area of a nail holding down a piece of plywood sheathing is very small, whereas the entire roof contributes load to the foundations.

Truss: A framework of members joined at their ends to form a rigid structure. The basic element of a truss is the triangle, with each member of the triangle designed to take either tensile or compressive forces (or both).

Vertical mode amplitude: Amplitude of a mode of vibration (for example, of a bridge deck) that involves predominantly vertical motion relative to the longitudinal axis of the bridge.

Vertical mode frequency: A natural frequency of a mode of vibration that involves predominantly vertical motion.

Vortex shedding: An aerodynamic phenomenon that occurs when a fluid flows past a nearly two-dimensional body (that is, one having a nearly constant cross section). The high-speed flow past the body interacts with the sheltered wake to produce an instability in the form of vortices rolling up on either side in alternating fashion. These rotating flows are then shed downstream from one side and then the other at a regular rate that is determined by the cross-sectional dimension of the body and the speed of the flow.

Wind energy: The energy contained in the fluctuating components of the wind.

BIBLIOGRAPHY

Bailey, T. J. 1990. *Toward a Canadian Program for the International Decade for Natural Disaster Reduction: A Report from the Joint Committee for the International Decade for Natural Disaster Reduction*. Ottawa: Canadian Academy of Engineering.

Barelli, M., J. White, and D. P. Billington. 2006. "History and Aesthetics of the Bronx-Whitestone Bridge." *Journal of Bridge Engineering* 11 (2): 230–40.

Bartlett, F. M., J. K. Galsworthy, D. Henderson, et al. 2007. "The Three Little Pigs Project: A New Test Facility for Full-Scale Small Buildings." In *Proceedings of the 12th International Conference on Wind Engineering, Cairns, Australia, July 2007*. CD-ROM.

Billington, D. P. 1985. *The Tower and the Bridge: The New Art of Structural Engineering*. Princeton, NJ: Princeton University Press.

Campbell, R. 1995. "Builder Faced Bigger Crisis Than Falling Windows." *Boston Globe*. March 3. http://www.pulitzer.org/archives/5826.

Canadian National Committee for the International Decade for Natural Disaster Reduction. 1994. *Canadian National Report, Prepared for the IDNDR Mid-Term Review and the 1994 World Conference on Natural Disaster Reduction, May 23–27, Yokohama, Japan*. Ottawa: Royal Society of Canada.

Cermak, J. E. 1975. "Evolution of Wind Engineering as a New Discipline: Comments from the Chairman." *WERC Newsletter* 1 (1): 2–3.

Chan, S. 2005. "A Bridge Too Fat." *New York Times*. February 18. http://www.nytimes.com/2005/02/18/nyregion/18bridge.html?scp=1&sq=a%20bridge%20too%20fat&st=cse.

Chen, P., and L. Robertson. 1972. "Human Perception Thresholds of Horizontal Motion." *Journal of the Structural Division, Proceedings of the American Society of Civil Engineers* 98:1681–95.

Davenport, A. G. 1960. *Wind Loads on Structures*. Technical Paper No. 88. NRCC-5576. Division of Building Research, National Research Council of Canada, Ottawa.

Davenport, A. G. 1961a. "The Application of Statistical Concepts to the Wind Loading of Structures." *Institution of Civil Engineers Proceedings* 19:449–72. doi:10.1680/iicep.1961.11304.

———. 1961b. "A Statistical Approach to the Treatment of Wind Loading on Tall Masts and Suspension Bridges." Department of Civil Engineering, University of Bristol, Bristol, England.

———. 1962a. "Buffeting of a Suspension Bridge by Storm Winds." *Journal of the Structural Division, American Society of Civil Engineers* 88:233–68.

———. 1962b. "The Response of Slender, Line-Like Structures to a Gusty Wind." *Institution of Civil Engineers Proceedings* 23:389–408.

———. 1965a. "The Relationship of Wind Structures to Wind Loading." In *Wind Effects on Buildings and Structures: Proceedings of the Conference Held at the National Physical Laboratory, Teddington, Middlesex, on 26th, 27th, and 28th June, 1963*, 53–111. London: Her Majesty's Stationery Office.

———. 1965b. "The Buffeting of Structures by Gusts." In *Wind Effects on Buildings and Structures: Proceedings of the Conference Held at the National Physical Laboratory, Teddington, Middlesex, on 26th, 27th, and 28th June, 1963*, 357–91. London: Her Majesty's Stationery Office.

———. 1969a. "The Boundary Layer Wind Tunnel Laboratory: The First Three Years." *Exponential: University of Western Ontario Engineering Journal* 3 (1): 3–9.

———. 1969b. *A Report on the Activities of the Boundary Layer Wind Tunnel Laboratory, November 1965 to March 1969*. Boundary Layer Wind Tunnel Laboratory, Faculty of Engineering Science, University of Western Ontario, London.

———. 1972. "An Approach to Human Comfort Criteria for Environmental Wind Conditions." Presented at the Colloquium on Building Climatology, Stockholm.

———. 1974. "The Use of Taut Strip Models in the Prediction of the Response of Long-Span Bridges to Turbulent Wind." In *Flow-Induced Structural Vibrations: Proceedings of the IUTAM-IAHR Symposium Held at Karlsruhe, Germany, August 14–16, 1972*, 373–82. Berlin: Springer-Verlag.

———. 1975. "The Design of Tall Buildings for Wind Forces: International Developments and the Present State of the Art." Presented at the *South African Conference on Tall Buildings*, Johannesburg.

———. 1977. "Wind Engineering, Ancient and Modern: The Relationship of Wind Engineering Research to Design." Presented at the Sixth Canadian Congress of Applied Mechanics, Vancouver.

———. 1979. "Sowing the Wind." Address on receiving an honorary degree from the University of Louvain, Belgium.

———. 1985. "The Role of Wind Engineering in Reducing the Risk of Windstorm Catastrophe." Presented at the First Asia-Pacific Symposium on Wind Engineering, Roorkee, India.

———. 1992a. "The Prediction of Hurricane Wind Speeds and Risk with Special Reference to the Caribbean." In *Proceedings of the Second Caribbean Conference on Fluid Dynamics, Faculty of Engineering, St. Augustine, Trinidad, January 5-8, 1992*. Saint Augustine: University of the West Indies.

———, ed. 1992b. *Proceedings of the Eighth International Conference on Wind Engineering, London, Ontario, Canada, July*. Amsterdam: Elsevier.

———. 1996. "Address to Carleton University Convocation." Ottawa: Carleton University.

———. 1999a. "The Missing Links." In *Wind Engineering into the 21st Century: Proceedings of the Tenth International Conference on Wind Engineering, Copenhagen/Denmark, 21–24 June 1999*, ed. A. Larsen, G. L. Larose, and F. M. Livesey, 3–14. London: A. A. Balkema.

———. 1999b. "The House That Won't Fall Down." In *Dealing with Natural Disasters: Achievements and New Challenges in Science, Technology and Engineering: Proceedings of a Conference Held 27–20 October 1999 at the Royal Society, London*. London: Royal Society. doiL 10.1.1.197.2443.pdf.

———. 2000a. "Application of Wind Engineering to the Changing World." In *Proceedings of the Third Structural Specialty Conference, Canadian Society for Structural Engineering Annual Conference, June 7-10, London, Ontario*. Montreal: Canadian Society for Structural Engineering. CD-ROM.

———. 2000b. "Comparison of Seismic and Windstorm Hazards." In *Proceedings of the Sixth Environmental Engineering Specialty Conference of the Canadian Society for Civil Engineering*, 504–9. Montreal: Canadian Society for Civil Engineering.

———. 2000c. "The Decade for Natural Disaster Reduction in Canada." *Natural Hazards Review* 1 (1): 27–36.

———. 2000d. "Remarks by Alan G. Davenport, Chair of the Canadian National Committee for the International Decade for Natural Disaster Reduction, to the Standing Senate Committee on National Finance." Ottawa.

———. 2000e. "The Role of Civil Engineering and Insurance in Mitigating Natural Disasters." In *Proceedings of the Canadian Society for Civil Engineering 2000 Conference, June 7-10, London, Ontario*. Montreal: Canadian Society for Civil Engineering. CD-ROM.

Davenport, A. G. 2001. "The World Trade Center Diary." Lecture to the Baconian Society, London.

———. 2002. "Past, Present and Future of Wind Engineering." *Journal of Wind Engineering and Industrial Aerodynamics* 90 (12–15): 1371–80.

Davenport, A. G., P. N. Georgiou, and D. Surry. 1986. "A Hurricane Risk Study for the Eastern Caribbean, Jamaica and Belize, with Special Considerations of the influences of Topography." In *Urban Climatology and Its Applications with Special Regard to Tropical Areas: Proceedings of the Technical Conference Organized by the world Meteorological Organization and Co-sponsored by the World Health Organization (Mexico D. F., 26–30 November 1984)*, ed. T. R. Oke. Geneva: WHO Secretariat.

Davenport, A. G., M. Hogan, and N. Isyumov. 1969. "A Study of Wind Effects on the Commerce Court. Part 1. Meteorological Study, Development of Structural Response Cladding Pressures." Research Report BLWT-7-69. Boundary Layer Wind Tunnel Laboratory, Faculty of Engineering Science, University of Western Ontario, London.

Davenport, A. G., M. Hogan, and B. J. Vickery. 1970. "An Analysis of Records of Wind Induced Building Movement and Column Strain Taken at the John Hancock Center." Research Report BLWT-10-70. Boundary Layer Wind Tunnel Laboratory, Faculty of Engineering Science, University of Western Ontario, London.

Davenport, A. G., E. Holm, and D. Surry. 1982. "The Definition of Steady and Fluctuating Wind Loads on Large Span Roofs." Presented at the Canadian Society for Civil Engineering Annual Conference, Edmonton.

Davenport, A. G., and N. Isyumov. 1982. "Further Studies of Wind Action for the Bronx-Whitestone Bridge in New York, NY." Research Report BLWT-SS17-82. Boundary Layer Wind Tunnel Laboratory, Faculty of Engineering Science, University of Western Ontario, London.

Davenport, A. G., N. Isyumov, and C. F. P. Bowen. 1970. "A Study of Wind Effect on the Commerce Court Project. Part 2. Wind Environment at Pedestrian Level." Research Report BLWT-3-70. Boundary Layer Wind Tunnel Laboratory, Faculty of Engineering Science, University of Western Ontario, London.

Davenport, A. G., N. Isyumov, D. F. Fader, and C. F. P. Bowen. 1970. "A Study of the Wind Effects on the World Trade Center, New York Exterior Pressures on Plaza Buildings and Airflow in Plaza." Research Report BLWT-6-70. Boundary Layer Wind Tunnel Laboratory, Faculty of Engineering Science, University of Western Ontario, London.

Davenport, A. G., N. Isyumov, and G. Greig. 1978. "A Study of Pedestrian Level Wind Environment Around the American Telephone and Telegraph

Corporate Headquarters and the International Business Machines Corporation Building, New York, NY." Research Report BLWTL-SS7-78. Boundary Layer Wind Tunnel Laboratory, Faculty of Engineering Science, University of Western Ontario, London.

Davenport, A. G., N. Isyumov, and T. Jandali. 1971. "A Study of Wind Effects for the Sears Project." Research Report BLWT-5-71. Boundary Layer Wind Tunnel Laboratory, Faculty of Engineering Science, University of Western Ontario, London.

Davenport, A. G., N. Isyumov, H. Rothman, and H. Tanaka. 1980. "Wind Induced Response of Suspension Bridges: Wind Tunnel Model and Full Scale Observations." In *Wind Engineering: Proceeding of the Fifth International Conference on Wind Effects on Buildings and Structures, Fort Collins, Colorado, USA, July 1979*, ed. J. E. Cermak, 807–24. Oxford: Pergamon Press.

Davenport, A. G., N. Isyumov, and H. Tanaka. 1976. "A Study of Wind Action for the Bronx-Whitestone Suspension Bridge, New York, NY." Research Report BLWT-SS3-76. Boundary Layer Wind Tunnel Laboratory, Faculty of Engineering Science, University of Western Ontario, London.

Davenport, A. G., and J. P. C. King. 1982a. "A Study of Wind Effects for the Sunshine Skyway Bridge, Tampa, Florida: Concrete Alternate." Research Report BLWT-SS24. Boundary Layer Wind Tunnel Laboratory, Faculty of Engineering Science, University of Western Ontario, London.

———. 1982b. "A Study of Wind Effects for the Sunshine Skyway Bridge, Tampa, Florida: Steel Alternate." Research Report BLWT-SS25-82. Boundary Layer Wind Tunnel Laboratory, Faculty of Engineering Science, University of Western Ontario, London.

———. 2000. "A Canadian Retrospective of Wind Engineering of Long Span Bridges." In *Proceedings of the Canadian Society for Civil Engineering 2000 Conference, June 7-10, London, Ontario*. Montreal: Canadian Society for Civil Engineering. CD-ROM.

Davenport, A. G., J. P. C. King, and G. Larose. 1992. "Taut Strip Model Tests." Presented at the International Symposium on Aerodynamics of Large Bridges, Copenhagen.

———. 1993. "Further Studies of Wind Effects for the Storebaelt Bridge, Denmark." Research Report BLWT-SS14-1993. Boundary Layer Wind Tunnel Laboratory, Faculty of Engineering Science, University of Western Ontario, London.

Davenport, A. G., Mackey, S., and Melbourne, W. H. 1980. "Wind Loading and Wind Effects." In Tall Building Criteria and Loading, Volume 150 of

Monographs on Tall Buildings Series, Council on Tall Buildings and Urban Habitat, ed. L .E. Robertson and E. H. Gaylord, 143–248. New York: American Society of Civil Engineers.

Davenport, A. G.; J. Monbaliu, P. Dance, and N. Isyumov. 1985. "Properties of the Atmospheric Boundary Layer Measured at the CN Tower, Toronto, Ontario, Canada." Presented at the Fifth National Conference on Wind Engineering, Lubbock, TX.

Davenport, A. G., and D. Surry. 1974. "The Pressures on Low-Rise Structures in Turbulent Wind." Presented at the Canadian Structural Engineering Conference, Toronto.

———. 1978. "A Study of Wind Effects on the Hajj Terminal at Jeddah International Airport: A Description of the Test Procedures for the Rigid Model Studies and a Summary of the Resulting Wind Loading Criteria." Research Report BLWT-SS9-78. Boundary Layer Wind Tunnel Laboratory, Faculty of Engineering Science, University of Western Ontario, London.

———. 1984. "Turbulent Wind Forces on a Large Span Roof and Their Representation by Equivalent Static Loads." *Canadian Journal of Civil Engineering* 11:955–66.

Davenport, A. G., D. Surry, and R. B. Kitchen. 1976. "Wind Induced Exterior Pressures on the John Hancock Tower, Boston." Research Report BLWT-SS2-76. Boundary Layer Wind Tunnel Laboratory, Faculty of Engineering Science, University of Western Ontario, London.

Davenport, A. G., D. Surry, and G. Lythe. 1984. "The Integration of Structural Analysis and Wind Tunnel Testing for the New Hongkong and Shanghai Banking Corporation Headquarters in Hong Kong." Presented at the Third International Conference on Tall Buildings, Hong Kong.

Davenport, A. G., D. Surry, T. Stathopoulas, and L Apperley. 1978. "Current Research on Wind Loads on Low-Rise Buildings." Presented at the *Third National Conference on Wind Engineering Research*, Gainesville, FL.

Davenport, A. G., D. Surry, and H. Tanaka, 1975. "A Study of the Wind Induced Response of the John Hancock Tower, Boston." Research Report BLWT-SS3-75. Boundary Layer Wind Tunnel Laboratory, Faculty of Engineering Science, University of Western Ontario, London.

Davenport, A. G., B. V. Tryggvason, and D. Surry. 1976. "Monte Carlo Simulation of Hurricanes for the Prediction of Wind Induced Response." Research Report BLWT-1-1976. Boundary Layer Wind Tunnel Laboratory, Faculty of Engineering Science, University of Western Ontario, London.

Davenport, Sheila. 1998. "Japanese Elevate Art of Bridge Building." *London Free Press*. May 16, G4.

D'Costa, M. J., and F. M. Bartlett. 2000. "Design of a Static-equivalent Wind-load Test for a Full-scale Corrugated Fibreboard Shelter." In *Proceedings of the Third Structural Specialty Conference*, Canadian Society for Structural Engineering Annual Conference, June 7–10, London, Ontario. Montreal: Canadian Society for Structural Engineering. CD-ROM.

———. 2003. "Full-Scale Testing of Corrugated Fibreboard Shelter Subjected to Static-Equivalent Wind Loads." *Journal of Wind Engineering & Industrial Aerodynamics* 91 (12–15): 1671–88.

De Villiers, Marq. 2006. *Windswept: The Story of Wind and Weather*. Toronto: McClelland and Stewart.

Economist. 1998. "The Bridge to Nowhere in Particular." April 4. http://www.economist.com/node/361147.

———. 2003. "A Bridge Too Far." November 20. http://www.economist.com/node/2235192.

Engineering News-Record. 1966. "Building Revolution Gains New Momentum." February 10, 100–102.

———. 1980. "Wind Analysis: Preventive Medicine for Cladding, Structural Problems. Codes Less Precise Than Wind Engineering." March 27, 26–30.

———. 1995. "Critics Grade Citicorp Confession." 234 (21): 10.

Fackler, Martin, 2011. "Tsunami Warnings, Written in Stone." *New York Times*. April 20. http://www.nytimes.com/2011/04/21/world/asia/21stones.html?_r=1&scp=1&sq=do%20not%20build&st=cse.

Finn, Robin, 2000. "Engineer Tracks Bridges' Twists and Turns." *New York Times*. December 13. http://www.nytimes.com/2000/12/15/nyregion/public-lives-engineer-tracks-bridges-twists-and-turns.html?scp=1&sq=Engineer%20Tracks%20Bridges%E2%80%99%20Twists%20and%20Turns&st=cse.

Gavanski, E., and G. A. Kopp. 2011a. "Glass Breakage Tests under Fluctuating Wind Loads." *ASCE Journal of Architectural Engineering* 17:34–41.

———. 2011b. "Examination of Load Resistance in Window Glass Design." *ASCE Journal of Architectural Engineering* 17:42–50.

———. 2011c. Storm and Gust Duration Effects on Design Wind Loads for Glass." *ASCE Journal of Structural Engineering* 137:1603–10.

Ge, Yaojun. 2002. "Professor A. G. Davenport and Wind Engineering in China." Presented at the Engineering Symposium to Honour Alan G. Davenport, University of Western Ontario, London. CD-ROM.

Georgiou, P. N., A. G. Davenport, and B. J. Vickery. 1983. "Design Wind Loads in Regions Dominated by Tropical Cyclones." *Journal of Wind Engineering and Industrial Aerodynamics* 13:139–52.

Georgiou, P. N., and B. J. Vickery. 1980. "Wind Loads on Building Frames." In *Wind Engineering: Proceedings of the Fifth International Conference on Wind Effects on Buildings and Structures, Fort Collins, Colorado, USA, July 1979*, ed. J. E. Cermak, 421–33. Oxford: Pergamon Press.

Glanz, James. 2003. Transcript of interview with Alan Davenport. March 20.

Glanz, James, and Eric Lipton. 2003. *City in the Sky: The Rise and Fall of the World Trade Center*. New York: Henry Holt.

Gibbs, Tony. 2002. "Professor Alan Davenport's Contributions to Wind Engineering in the Caribbean." Presented at the Engineering Symposium to Honour Alan G. Davenport, University of Western Ontario, London. CD-ROM.

Hangan, Horia. 2008. "The Wind Engineering, Energy and Environment Dome—Project Module, New Initiatives Fund, Canada Foundation for Innovation." Boundary Layer Wind Tunnel Laboratory, Faculty of Engineering Science, University of Western Ontario, London.

———. 2010. "Current and Future Directions for Wind Research at Western: A New Quantum Leap in Wind Research through the Wind Engineering, Energy and Environment (WindEEE) Dome." *Japan Architectural Wind Engineering Journal* 35 (4): 277–81.

Henderson, D. J., J. D. Ginger, M. J. Morrison, and G. A. Kopp. 20 09. "Simulated Tropical Cyclonic Winds for Low Cycle Fatigue Loading of Steel Roofing." *Wind and Structures* 12 (4): 381–98.

Horsley, Carter B. 1974. "The Wind, Fickle and Shifty, Tests Builders." *New York Times*. May 5. http://select.nytimes.com/gst/abstract.html?res=F306 16F73B5F107A93C7A9178ED85F408785F9&scp=2&sq=alan%20g.%20 Davenport&st=cse.

Huler, Scott. 2004. *Defining the Wind: The Beaufort Scale and How a 19th-Century Admiral Turned Science into Poetry*. New York: Crown.

International Association for Earthquake Engineering. 1992. *Proceedings of the Tenth World Conference on Earthquake Engineering, 19–24 July, Madrid, Spain*. Rotterdam: A. A. Balkema.

International Council for Science. 2008. "A Science Plan for Integrated Research on Disaster Risk." Draft. http://www.icsu.org/publications/reports-and-reviews/IRDR-science-plan.

Isyumov, N. 2011. "Alan G. Davenport's Mark on Wind Engineering." In *Proceedings of the 13th International Conference on Wind Engineering, Amsterdam*. Amsterdam: ICWE. CD-ROM.

Isyumov, N., and A. G. Davenport. 1975a. "The Ground Level Wind Environment in Built-Up Areas." In *Proceedings of the Fourth International*

Conference on Wind Effects on Buildings and Structures, ed. Keith John Eaton, 403–19. Heathrow: Cambridge University Press.

Isyumov, N., and A. G. Davenport. 1975b. "A Study of Wind-Induced Exterior Pressures and Suctions on Lower Accommodation Levels of the CN Tower." Research Report BLWT-SS2-75. Boundary Layer Wind Tunnel Laboratory, Faculty of Engineering Science, University of Western Ontario, London.

Isyumov, N., A. G. Davenport, and J. Monbaliu. 1984. "CN Tower, Toronto: Model and Full Scale Response to Wind." Presented at the 12th Congress of the International Association for Bridge and Structural Engineering, Vancouver.

Isyumov, N., and R. A. Halvorson. 1984. "Dynamic Response of Allied Bank Plaza during Alicia." In *Proceedings of ASCE Specialty Conference, "Hurricane Alicia: One Year Later,"* ed. Ahsan Kareem, 98–116. New York: American Society of Civil Engineers.

Isyumov, N., F. Knoll, J. Mardukhi, and D. P. Morrish. 2000. "CN-Tower, Revisit of Antenna Performance under Wind Action." In *Proceedings of the Canadian Society for Civil Engineering 2000 Conference, June 7–10, London, Ontario.* Montreal: Canadian Society for Civil Engineering. CD-ROM.

Jensen, M. 1958. "The Model Law for Phenomena in the Natural Wind." *Ingeniören* 2:121–28.

Kashima, S., and M. Kitagawa. 1997. "The Longest Suspension Bridge." *Scientific American.* December. http://www.sciamdigital.com/index.cfm?fa =Products.ViewIssuePreview&ISSUEID_CHAR=449B9270-2B35 -221B-6F633433EF28DC14&ARTICLEID_CHAR=44A3C0CC-2B35 -221B-647208FC23C7C620.

Khan, Y. S. 2004. *Engineering Architecture: The Visions of Fazlur R. Khan.* New York: W. W. Norton.

King, J. P. C. 2003. "The Foundation and Future of Wind Engineering of Long Span bridges: The Contributions of Alan Davenport." *Journal of Wind Engineering and Industrial Aerodynamics* 91 (12–15): 1529–46.

King, J. P. C., A. G. Davenport, and G. Larose. 1991. "A Study of Wind Effects for the Storebaelt Bridge, Tender Design, Denmark." Research Report BLWT-SS31-1991. Boundary Layer Wind Tunnel Laboratory, Faculty of Engineering Science, University of Western Ontario, London.

King, J. P. C., T. C. E. Ho, M. J. Mikitiuk, and A. G. Davenport. 1993. "Lantau Fixed Crossing—Tsing Ma Bridge, Deck Erection Aerodynamic Investigation, Final Report." Research Report BLWT-SS23-1993. Boundary

Layer Wind Tunnel Laboratory, Faculty of Engineering Science, University of Western Ontario, London.

King, J. P. C., G. A. Kopp, L. Z. Kong, and N. Isyumov. 2000. "A Study of Wind Effects for The Bronx-Whitestone Bridge—Task 1B4—Full Aeroelastic Model Investigation." Research Report BLWT-SS31-2000. Boundary Layer Wind Tunnel Laboratory, Faculty of Engineering Science, University of Western Ontario, London.

King, J. P. C., G. L. Larose, and A. G. Davenport. 1991. "A Study of Wind Effects for the Paso Del Alamillo Bridge, Sevilla, Spain." Research Report BLWT-SS28-1991. Boundary Layer Wind Tunnel Laboratory, Faculty of Engineering Science, University of Western Ontario, London.

King, J. P. C., D. P. Morrish, N. Isyumov, and G.-N. Fanjiang. 2000. "Full-scale Monitoring of the Bronx-Whitestone Bridge." In *Proceedings of the Canadian Society for Civil Engineering 2000 Conference, June 7-10, London, Ontario*. Montreal: Canadian Society for Civil Engineering. CD-ROM.

Kopp, G. A., and E. Gavanski. 2012. "Effects of Pressure Equalization on the Performance of Residential Wall Systems under Extreme wind Loads." *ASCE Journal of Structural Engineering* 138 (4). doi: 10.1061/(ASCE) ST.1943-541X.0000476.

Kopp, G. A., M. J. Morrison, E. Gavanski, et al. 2010. "'Three Little Pigs' Project: Hurricane Risk Mitigation by Integrated Wind Tunnel and Full-Scale Laboratory Tests." *ASCE Natural Hazards Review* 11 (4): 151–61.

Kopp, G. A., M. J. Morrison, and David J. Henderson. 2011. "Full-Scale Testing of Low-Rise Buildings Using Realistically Simulated Wind Loads." In *Proceedings of the 13th International Conference on Wind Engineering, Amsterdam*. Amsterdam: ICWE. CD-ROM.

Kopp, G. A., M. J. Morrison, B. Kordi, and C. Miller. 2011. "A Method to Assess Peak Storm Wind Speeds Using Detailed Damage Surveys." *Engineering Structures* 33 (1): 90–98.

Krishna, P., ed. 1995. A State of the Art in Wind Engineering. In *Proceedings, International Conference on Wind Engineering, Davenport Sixtieth Birth Anniversary Volume*. New Delhi: Wiley Eastern.

———. 2002. "Professor Alan Davenport and India." Presented at the Engineering Symposium to Honour Alan G. Davenport, University of Western Ontario, London. CD-ROM.

Larose, G., A. G. Davenport, and J. P. C. King. 1992. "Wind Effects on Long Span Bridges: Consistency of Wind Tunnel Results." *Journal of Wind Engineering and Industrial Aerodynamics* 42 (1–3): 1191–1202.

Le Temps. 1887. "Protest against the Tower of Monsieur Eiffel." http://www

.eiffel-tower.com/component/content/article/16-dossiers-thematiques/ 71-debats-et-polemiques-a-lepoque-de-la-construction.html.

Levy, M., and M. Salvadori, 1992. *Why Buildings Fall Down, How Structures Fail.* New York: W. W. Norton.

Lythe, G., and D. Surry, 1982. "An Experimental Study of the Wind Environment in Areas Around the New Hongkong and Shanghai Bank Building, Hong Kong." Research Report BLWT-SS11-1982. Boundary Layer Wind Tunnel Laboratory, Faculty of Engineering Science, University of Western Ontario, London.

Mardukhi, J., N. Isyumov, and F. Knoll. 2001. "CN-Tower Performance Over Half a Century." In *Proceedings of the Canadian Society for Civil Engineering 29th Annual Conference, May 30 to June 2, 2001, Victoria, BC.* Montreal: CSCE.

Metals in Construction. 2006. "Bronx-Whitestone Bridge Re-Decking." Fall, 38–41.

Miller, N., and J. Murphy. 1980. "Thinking Tall." *Progressive Architectures,* December, 45–57.

Milne, W. G., and A. G. Davenport. 1969. "Distribution of Earthquake Risk in Canada." *Bulletin of the Seismic Society of America* 59 (2): 729–54.

Miyata, T., and M. Yasuda. 1993. "Full Model Wind Tunnel Tests for the Akashi Kaikyo Bridge." In *Structural Engineering in Natural Hazards Mitigation: Proceedings of Papers Presented at the Structures Congress '93, Irvine, California, April 19–21, 1993,* 490–95. New York: American Society of Civil Engineers.

Morgenstern, Joe. 1995. "The Fifty-Nine-Story Crisis." *New Yorker,* May, 45–53.

Moses, Robert. 1970. *Public Works: A Dangerous Trade.* New York: McGraw-Hill.

National Geographic. 1998. "Natural Hazards of North America/Great Disasters, Nature in Full Force." July. Insert.

National Physical Laboratory. 1965. *Wind Effects on Buildings and Structures: Proceedings of the Conference held at the National Physical Laboratory, Teddington, Middlesex, on 26th, 27th, and 28th June, 1963.* London: Her Majesty's Stationery Office.

National Public Radio. 1995. "U.N. Sponsored Natural Disaster Reduction Day." *All Things Considered.* October 11. Transcript.

New York Times. 1939. "Whitestone Span Opened by Mayor." April 30. http://select.nytimes.com/gst/abstract.html?res=F60716FA3A5A167A93 C2AA178FD85F4D8385F9&scp=3&sq=Whitestone%20Span%20 Opened%20by%20Mayor&st=cse.

New York Times. 1940a. "Big Tacoma Bridge Crashes 190 Feet into Puget Sound." November 7. http://select.nytimes.com/gst/abstract.html?res=F3 0911FC3C5C10728DDDA10894D9415B8088F1D3&scp=1&sq=Big%20 Tacoma%20Bridge%20Crashes%20190%20Feet%20into%20Puget%20 Sound&st=cse.

———. 1940b. "Cables to End Sway in Whitestone Bridge; Moses Denies Tacoma Crash Caused Change." December 5. http://select.nytimes.com/ gst/abstract.html?res=FB0611F73558127A93C7A91789D95F448485F9 &scp=3&sq=Cables%20to%20End%20Sway%20in%20Whitestone%20 Bridge&st=cse.

Ontario Climate Centre. 1996. "WWCN13 CWTO 202242: Tornado Warning for Regions of Southern Ontario." April 20. Toronto: Environment Canada.

Palmer-Benson, T. 1987. "Wind Breakers: Building Design with Pedestrian Comfort in Mind." *Canadian Geographic,* February/March, 34–37.

Perlmutter, E. 1968. "Bridge and Ferry Users Have Rough Crossing." *New York Times.* November 13. http://select.nytimes.com/gst/abstract.html?res =F00A14FE3954157493C1A8178AD95F4C8685F9&scp=1&sq=perlmut ter%20Bridge%20and%20Ferry%20Users%20Have%20Rough%20 Crossing&st=cse.

Petroski, H. 1985. *To Engineer Is Human, The Role of Failure in Successful Design.* New York: St. Martin's Press.

———. 1991. "Still Twisting." *American Scientist* 79:398–401.

——— 1998. "New and Future Bridges." *American Scientist* 86:514–18.

Pollalis, S. 1999. *What Is a Bridge? The Making of Calatrava's Bridge in Seville.* Cambridge, MA: MIT Press.

Pugsley, A. 1957. *The Theory of Suspension Bridges.* London: Edward Arnold.

Rastorfer, D. 2000. *Six Bridges: The Legacy of Othmar H. Ammann.* New Haven, CT: Yale University Press.

Revkin, A., 2008. "On Elephants' Memories, Human Forgetfulness and Disaster." *New York Times.* April 12. http://dotearth.blogs.nytimes.com/2008/ 08/12/on-elephants-memories-human-forgetfulness-and-disaster/.

Roberts, S. 2007. "It Will Take More Than a Wolf to Blow One House Down." *New York Times.* July 3.

Roth, A. 2003. "A Onetime Thing of Beauty Gets a Little Prettying Up." *New York Times.* October 12. http://www.nytimes.com/2003/10/12/nyregion/ neighborhood-report-whitestone-onetime-thing-beauty-gets-little -prettying-up.html?scp=1&sq=A%20Onetime%20Thing%20of%20 Beauty%20Gets%20a%20Little%20Prettying%20Up&st=cse.

Rothman, H. 1995. "Aerodynamic Stabilization of the Bronx-Whitestone Bridge." In *Restructuring: America and Beyond, Proceedings of Structures Congress XIII, Boston, Massachusetts, April 2–5, 1995*, ed. Masoud Sanayei, 1554–57. New York: American Society of Civil Engineers.

Royal Society of Canada. 1993. "News Release, October 13: International Day for Natural Disaster Reduction." Ottawa: Royal Society of Canada.

Salvadori, M. 1990. *Why Buildings Stand Up: The Strength of Architecture.* New York: W. W. Norton.

Sayers, A. 2008. "Critical Analysis of Sunshine Skyway Bridge." In *Proceedings of Bridge Engineering 2 Conference, May 4, 2007, University of Bath, Bath UK.* https://docs.google.com/viewer?a=v&q=cache:Sdu4kr1dDUQJ :www.bath.ac.uk/ace/uploads/StudentProjects/Bridgeconference2007/ conference/mainpage/Sayers_Sunshine_Skyway.pdf+Critical+Analysis +of+Sunshine+Skyway+Bridge&hl=en&gl=ca&pid=bl&srcid=ADGE ESi_wWDSOc-YfGgyf10Qqa8fx15q9BFssDhCNh3KfjK2ih2o ZBtj0EuC2EqhRY0CN3U3GRsYK-zodHdwKTAV54EUhcMg1c PLq0xxYi060NvxG9vNn-npeOgxUbLBxtQFb1-GT3HB&sig=AHIEtb RW6L8Qjg1tXpsdvGC-q7Jbh1a6rA.

Scanlan, R., and E. Simiu. 1996. *Wind Effects on Structures: Fundamentals and Applications to Design.* New York: Wiley-Interscience.

Scanlan, R., and K. Yusuf Billah. 1991. "Resonance, Tacoma Narrows Bridge Failure, and Undergraduate Physics Textbooks." *American Journal of Physics* 59 (2): 118–24.

Schuyler, M. 1909. "The Evolution of the Skyscraper." *Scribner's Magazine*, September, 257–71.

Scott, R. 2001. *In The Wake of Tacoma: Suspension Bridges and the Quest for Aerodynamic Stability.* Reston, VA: American Society of Civil Engineers.

Scruton, K. 1981. *An Introduction to Wind Effects on Structures.* Oxford: Oxford University Press.

Seabrook, J. 2001. "The Tower Builder." *New Yorker.* November 19, 65–73.

Seelye, K. Q. 2011. "Year Packed with Weather Disasters Has Brought Economic Toll to Match." *New York Times.* August 20. http://www.nytimes. com/2011/08/20/us/20weather.html?hp.

Shipnuk, Alan. 2002. "Tunnel Vision." *Sports Illustrated.* April 1998. [SB: resolve inconsistency in dates?] http://sportsillustrated.cnn.com/vault/article/magazine/MAG1025488/index.htm

Stathopoulos, T., D. Surry, and A. G. Davenport. 1980. "Internal Pressure Characteristics of Low-Rise Buildings Due to Wind Action." In *Wind Engineering: Proceedings of the Fifth International Conference on Wind Effects* on *Buildings and Structures, Fort Collins, Colorado, USA, July 1979*, ed. J. E. Cermak, 451–63. Oxford: Pergamon Press.

Sullivan, Louis. 1896. "The Tall Office Building Artistically Considered." *Lippincott's Magazine* 57:403–9.

Surry, D., G. Lythe, and A. G. Davenport. 1981. "Wind-Induced Exterior Suctions and Pressures on the New Hong Kong and Shanghai Bank Building, Hong Kong." Research Report BLWT-SS14-81. Boundary Layer Wind Tunnel Laboratory, Faculty of Engineering Science, University of Western Ontario, London.

———. 1983. "Wind-Induced Overall Loads and Accelerations on the New Hong Kong and Shanghai Bank Building, Hong Kong." Research Report BLWT-SS7-83. Boundary Layer Wind Tunnel Laboratory, Faculty of Engineering Science, University of Western Ontario, London.

Surry, D., D. Meecham, E. M. F. Stopar, and M. Cholod. 1989. "The Development of Models Designed to Fail under Wind Loading." *Journal of Wind Engineering and Industrial Aerodynamics* 32 (3): 343–60.

Surry, D., and T. Stathopoulos. 1978. "An Experimental Approach to the Economical Measurement of Spatially Averaged Wind Loads." *Journal of Wind Engineering and Industrial Aerodynamics* 2 (4): 385–97.

Surry, D., H. Tanaka, K. N. Allen, and A. G. Davenport. 1975. "Investigation of the Wind Speeds Indicated By the Anemometer on the John Hancock Tower and Their Ratio to the Gradient Wind Speed." Research Report BLWT-SS4-1975, Boundary Layer Wind Tunnel Laboratory, Faculty of Engineering Science, University of Western Ontario, London.

Talese, G. 1964. "City Bridge Creator, 85, Keeps Watchful Eye on His Landmarks: O.H. Ammann Views 8 Spans by Telescope From Suite in Carlyle 32 Stories Up." *New York Times*. March 26. http://select.nytimes.com/gst/abstract.html?res=F00612F6395415738DDDAF0A94DB405B848AF1D3&scp=2&sq=Keeps%20Watchful%20Eye%20on%20His%20Landmarks&st=cse.

Thürlimann, B. 1975. *John Hancock Tower, Boston: Structural Performance and Safety Margins: Report prepared for I.M. Pei and Partners.* Report 7403-2. Zurich: Swiss Federal Institute of Technology.

Tryggvason, B. V., A. G. Davenport, and D. Surry. 1976. "Predicting Wind-Induced Response in Hurricane Zones." *ASCE Journal of the Structural Division* 102 (12): 2333–50.

Tryggvason, B. V., D. Surry, and A. G. Davenport. 1979. "A Study of Wind Effects on an Aeroelastic Model of the Hajj Terminal at Jeddah International Airport: A Summary Report of the Test Procedures, Observations and Conclusions." Boundary Layer Wind Tunnel Laboratory, Faculty of Engineering Science, University of Western Ontario, London.

Tschanz, T., and A. G. Davenport. 1983. "The Balance Technique for the Determination of Dynamic Wind Loads." *Journal of Wind Engineering and Industrial Aerodynamics* 13 (1–3): 429–39.

United Nations. 1989. "General Assembly, A/RES/44/236, 85th Plenary Meeting, 22 December 1989, International Decade for Natural Disaster Reduction." http://www.un.org/documents/ga/res/44/a44r236.htm.

———. 1999. "Press Release, SG/SM/7060, Secretary-General to Decade for Natural Disaster Reduction: Despite Dedicated Efforts, Number and Cost of Natural Disasters Continue to Rise." http://www.un.org/News/Press/docs/1999/19990706.SGSM7060.html.

Velez, G., and G. Fanjiang. 2005. "Bronx-Whitestone Bridge Truss Removal and Installation of Aerodynamic Wind Fairings." Presented at the Bruce Podwal Seminar Series, Department of Civil Engineering, City College of New York.

Weidman, P., and I. Pinelis. 2004. "Model Equations for the Eiffel Tower Profile: Historical Perspective and New Results." *Comptes Rendus Mecanique* 332:571–84.

Worthington, Skilling, Helle & Jackson. 1966. *The World Trade Center Wind Study: Final Report.* New York: Worthington, Skilling, Helle & Jackson.

Wells, M. 2002. *30 Bridges.* New York: Watson-Guptill.

White, R. 1992. Audio interview with Alan Davenport. Canadian Society for Civil Engineering Oral History Project, London.

Williams, S. 1989. *Hongkong Bank: The Building of Norman Foster's Masterpiece. New York:* Little, Brown.

INDEX

A page number in italics refers to an illustration or its caption.